EMERGENT EVOLUTION

Episteme

A SERIES IN THE FOUNDATIONAL,
METHODOLOGICAL, PHILOSOPHICAL, PSYCHOLOGICAL, SOCIOLOGICAL, AND POLITICAL ASPECTS OF THE SCIENCES, PURE AND APPLIED

VOLUME 19

The titles published in this series are listed at the end of this volume.

EMERGENT EVOLUTION

Qualitative Novelty and the Levels of Reality

by

DAVID BLITZ

Central Connecticut State University, U.S.A.

KLUWER ACADEMIC PUBLISHERS

DORDRECHT / BOSTON / LONDON

Library of Congress Cataloging-in-Publication Data

Blitz, David.
Emergent evolution : qualitative novelty and the levels of reality
David Blitz.
p. cm. -- (Episteme ; v. 19)
Includes bibliographical references (p.) and index.
ISBN 0-7923-1658-4 (alk. paper)
1. Evolution--History. 2. Evolution (Biology)--History.
I. Title. II. Series: Episteme (Dordrecht, Netherlands) ; v. 19.
B818.B65 1992
116--dc20 92-4950

ISBN 0-7923-1658-4

Published by Kluwer Academic Publishers,
P.O. Box 17, 3300 AA Dordrecht, The Netherlands.

Kluwer Academic Publishers incorporates
the publishing programmes of
D. Reidel, Martinus Nijhoff, Dr W. Junk and MTP Press.

Sold and distributed in the U.S.A. and Canada
by Kluwer Academic Publishers,
101 Philip Drive, Norwell, MA 02061, U.S.A.

In all other countries, sold and distributed
by Kluwer Academic Publishers Group,
P.O. Box 322, 3300 AH Dordrecht, The Netherlands.

Printed on acid-free paper

Printed in the Netherlands

TABLE OF CONTENTS

Part Two
The Origin of Emergent Evolution

Part Three
The Debate Over Emergent Evolution

ACKNOWLEDGEMENTS

I would like to acknowledge the encouragement of colleagues and friends, though none, of course, are responsible for errors or omissions: Mario Bunge, William Shea, Michael Hallett, Marguerite Deslauriers, and David Norton (McGill University: Montreal, Quebec), Pablo Iannone, Scott Paeth, Karen Beyard, and David Gerstein (Central Connecticut State University: New Britain, Connecticut), Daniel Fischlin (Bishop's University: Lennoxville, Quebec), Surindar Paracer (Worcester State College and Clark University: Worcester, Massachusetts), Richard Pattenaude (University of Southern Maine: Portland, Maine), Mike Dillinger (Federal University of Minas Gerais, Brazil), Guy Joseph (ConnecTalk, Montreal), Manfred Kory (translator, Montreal), and Adrian King-Edwards (Word Bookstore, Montreal). Research at the Lloyd Morgan Archives at the University of Bristol was aided by a CCSU/AAUP grant (American Association of University Professors), and earlier research was supported by FCAR/Québec fellowships (Fonds d'aide aux chercheurs et à la recherche). I also would like to thank Nicholas Lee of the University of Bristol Library for his archival assistance, and Annie Kuipers of Kluwer Academic Publishers for her editorial advice. This book is dedicated to the memory of my father, Morty Blitz.

Quotes used in this book came from many sources, including works produced by the following publishers: Cambridge University Press, Charles C. Thomas Publisher, Clark University Press, Columbia University Press, D. Appleton and Co., D. Reidel Publishing Co. (Kluwer Academic Publishers), Edward Stanford, George Allen and Unwin, Green and Co., Harvard University Press, Henry Holt and Co., J. Cattell Press, John Bale, John W. Parker and Son, Johns Hopkins Press, Longmans and Co., Macmillan and Co., Methuen and Co., Open Court Co., Oxford University Press, P. F. Collier and Son, Pergmanon Press, Princeton University Press, Routledge and Kegan Paul, Rowman and Littfield, The Pilot Press, Truebner and Co., University of California Press, University of Chicago Press, University of Minnesota Press, W. W. Norton and Co., and Williams and Norgate.

The following journals have been quoted from: British Journal for the Philosophy of Science, Dialectica, International Journal of General Systems, Journal of Philosophical Studies, Journal of Philosophy, Monist, Nature, Neuroscience, Philosophical Review, Psychological Review, Review of Metaphysics, and Science.

This book was set in Times-Roman 10/12 point type. The original text was prepared on a Macintosh LC computer, using Microsoft Word 4.0 and MacDraw II, and was output using an Apple Personal LaserWriter NT printer.

INTRODUCTION

Emergent evolution combines three separate but related claims, whose background, origin, and development I trace in this work: firstly, that evolution is a universal process of change, one which is productive of qualitative novelties; secondly, that qualitative novelty is the emergence in a system of a property not possessed by any of its parts; and thirdly, that reality can be analyzed into levels, each consisting of systems characterized by significant emergent properties.

In part one I consider the background to emergence in the 19th century discussion of the philosophy of evolution among its leading exponents in England – Charles Darwin, Herbert Spencer, T. H. Huxley, Alfred Russel Wallace, and G. J. Romanes. Unlike the scientific aspect of the debate which aimed to determine the factors and causal mechanism of biological evolution, this aspect of the debate centered on more general problems which form what I call the "philosophical framework for evolutionary theory." This considers the status of continuity and discontinuity in evolution, the role of qualitative and quantitative factors in change, the relation between the organic and the inorganic, the relation between the natural and the supernatural, the mind-body problem, and the scope of evolution, including its extension to ethics and morals.

Darwin argued that evolution proceeded in a continuous and gradual way, which he summed up in the aphorism "nature does not make jumps." Moreover, he believed that the continuous changes which occurred in nature were of a quantitative sort: structures increased or decreased in size, and were combined into further structures as the result of the action of natural selection. This produced populations of transformed individuals which were classified by biologists as new species. Differences of quantity, rather than differences of kind, characterized evolutionary change. Given Darwin's assumptions of the continuity of the evolutionary process and the quantitative nature of change, he concluded that mind, since present in humans, must have been present in all earlier forms of life, though to a lesser degree. Darwin's position of evolutionary continuity and quantitative change, taken by Ernst Haeckel to its logical extreme, meant attributing some form of psychic or mental qualities even to inanimate objects – panpsychism.

On the other hand, Alfred Russel Wallace, the co-discoverer of natural selection, argued that qualitative novelties did exist, in particular those of life and mind. Though he thus avoided attributing embryonic consciousness to amoebas and atoms, Wallace, a believer in the supernatural, thought that the source of the qualitative novelties of life and mind was to be found outside nature. Evolutionary philosophy appeared to be caught on the horns of a dilemma: either panpsychism or supernaturalism. This posed the question that emergent evolution would answer:

Is it possible to combine recognition of qualitative novelty in the evolutionary process, as proposed by Wallace, with a continuous and naturalistic evolutionary process, as required by Darwin?

In part two, I examine the work of Conwy Lloyd Morgan, the comparative psychologist who developed the philosophy of emergent evolution to address this problem. Lloyd Morgan's theory of emergent evolution was the result of a "creative synthesis" of many influences, placing emergence for the first time at the center of a philosophy of evolution. As a synthesis, emergent evolution combined a number of components. These included the monistic philosophical framework defended by Spencer, Huxley, and Romanes, the concept of emergence developed by J. S. Mill and G. H. Lewes, as well as Henri Bergson's notion of the creativity of evolution, E. G. Spaulding's theory of the analysis of wholes into parts, and Walter Marvin's concept of the levels of reality. Lloyd Morgan's originality lay in placing the idea of emergence at the heart of a philosophy of evolution, and in his theory of the material, biological, and psychological levels of reality. Though Lloyd Morgan's philosophy was not without its conceptual problems, it set the themes for the ensuing debate over emergence, qualitative novelty, and the levels of reality.

In part three I examine the debate over emergent evolution and argue that it can be characterized as a philosophical trend, which has gone through three phases: its "emergence" and the initial debate surrounding it during the period 1915 - 1935, its "eclipse" relative to reductionism and physicalism from the mid-1930s to the early 1950s, and its subsequent "re-emergence" as a viable concept in philosophy of science and systematic philosophy based on science. Just as a number of influences merged in the creative synthesis of Lloyd Morgan, so his theory of emergent evolution diverged into a multitude of interpretations and developments. I have chosen from this rich fund of theories a representative sampling, laying particular emphasis on those which included theories of the levels of reality.

I conclude with an evaluation of issues raised in the history of emergent evolution: the part-whole relation, qualitative novelty, and the levels of reality, and present my own proposal for a version of emergent evolution. This contribution is necessarily tentative and subject to correction, for the history of emergent evolution shows that theories evolve over time; they never come to rest in any definitive formulation, and must always be reconsidered in the light of criticisms and new ideas.

PART ONE

THE BACKGROUND TO EMERGENT EVOLUTION

CHAPTER ONE

Charles Darwin: The Framework for Evolutionary Theory

The scientific aspect of the debate over evolution centers on questions concerning its mechanism: Is natural selection the sole factor responsible for the origin and transformation of species, or are other factors required as well? What is the genetic basis for evolution? What is the rate of species formation - does evolution proceed gradually or does it admit of rapid speciation as well? These and related scientific questions do not exhaust the subject of evolution. The scientific theory of evolution is informed by philosophy, which serves to clarify its theoretical framework by analyzing concepts – such as quantity and quality, continuity and novelty, and substance and change – which are essential to it, and which it shares with other sciences. What I term the "philosophical framework for evolutionary theory" replies to questions including the following: Is the evolutionary process continuous or discontinuous? Is evolution exclusively quantitative or does it admit of qualitative change as well? Is evolution a purely natural process, or does it also require assistance from a divine source? Is the scope of evolution limited to the biological or is it universal, operating in all domains of reality?

In order to consider these questions, I present the basic principles of Charles Darwin's theory of evolution through natural selection, and then examine the philosophical framework he provided for his own scientific theory. This not only set the issues for the 19th century discussion with other theorists, particularly Herbert Spencer, T. H. Huxley, Alfred Russel Wallace, and George Romanes, but also posed the question the comparative psychologist Conwy Lloyd Morgan would attempt to answer, at the beginning of the 20th century, with the philosophy of emergent evolution.

(a) Natural Selection and Descent with Modification

Charles Darwin (1809-1882) in *Origin of Species* (1859, 6th edition 1872) set evolution on a scientific basis for the first time, succeeding where Jean Baptiste Lamarck (1744-1829) in *Zoological Philosophy* (1809) of half a century earlier had failed.[1] In Darwin's theory, the scientific mechanism of evolution included natural selection, as the chief but not the sole factor of species transformation, sexual selection, an adjunct to natural selection but distinct from it, and the inheritance of acquired characteristics, based on the putative hereditary mechanism of pangenesis.[2]

Darwin began *Origin of Species* not with a general statement of his theory, but with a chapter on variation in pigeons and domesticated animals. He was setting the stage for his theory of evolution with an analysis of the basic condition for that process: the fact that in nature, individuals in a population vary from the mean or the norm. Darwin admitted "our ignorance of the cause of each particular variation," (1872, 1: 190) and considered that the overall result appeared as if due to chance.[3] These variations were slight, as Darwin believed that "sudden and considerable" deviations of structure, especially in the wild, were monstrosities fatal to the individual, and therefore of no use to the species. Whatever the factors of variation might be, the result was that individuals of the same species were marked by small, but ultimately significant differences.

Individual variation in itself furnished only the context or the conditions for the operation of further components of the mechanism of evolution. Population increased geometrically, and if unchecked, the progeny of a single species would rapidly populate the planet. But potentially unrestricted population growth was accompanied by actually limited natural resources, and the result was a population too great for the available means of support, a thesis formulated in a different context by Thomas Malthus (1766-1834).[4] As the result of overpopulation, a struggle for existence ensued, which Darwin described as follows: "All that we can do is to keep steadily in mind that each organic being is striving to increase in a geometric ratio; that each at some period of its life, during some season of the year, during each generation or at intervals, has to struggle for life and to suffer great destruction." (1872, 1: 118-119)

This struggle for existence was resolved through natural selection which acted to preserve those individuals whose chance endowment of characteristics best suited them to meet the needs of their environment, and eliminated those whose characteristics were insufficient for the task at hand: "It may metaphorically be said that natural selection is daily and hourly scrutinizing, throughout the world, the slightest variations; rejecting those that are bad, preserving and adding up all that are good; silently and insensibly working, whenever and wherever opportunity offers, at the improvement of each organic being in relation to its organic and inorganic conditions of life." (1872, 1: 126)

Because of individual differences, members of a population possessed slightly different amounts of any given characteristic, including those required for survival in their immediate environment. Those individuals which were endowed with a greater amount of characteristics which enabled them to procure food and other necessaries of life tended to more successful than individuals which possessed those same characteristics to a lesser degree.[5] Natural selection, as a relation of correspondence between the capabilities of an organism and the requirements of its environment, therefore resulted in one of two outcomes, the first a positive and preserving one, the second a negative and destructive one. Organisms which were able to meet the environmental challenge and compete with their co-inhabitants tended to survive and reproduce; those which did not tended to die or failed to leave progeny.

Darwin conceived of natural selection as analogous to artificial selection. Through artificial selection, within the limits set by external conditions and the laws of heredity, humans could select for characteristics and develop breeds which met their needs and fancies. Just as artificial selection among domesticated animals produced new breeds, so natural selection in the wild produced new species. The analogy only extended so far, however, as there was an animate agent in artificial selection in the person of the breeder, while there was none, except in a very metaphorical sense, in natural selection. In addition, the time required for the production of breeds was much less than the time required for the transformation of species.

The above model assumed a stable environment, and concerned just one generation, while the full Darwinian theory took into account the effect of a changing environment, and the positive and negative effects of natural selection over many generations. In the negative and destructive case, natural selection resulted in the extinction of groups of individuals unsuited for the changes in their environment, while in the positive and creative case, there was a sufficient number of individuals possessing chance variations to take advantage of their modified circumstances. Characteristics not important in the old environment were selected for in the new, and after the process had been repeated for many thousands of generations in a changing environment, the resulting population differed significantly from the original population, or in Darwin's terms, had diverged from it.

Darwin compared the biological principle of the divergence of character to that of the economic division of labor, following in this the lead of H. Milne-Edwards (1800-1885).[6] In *Natural Selection* (1856-58, in Darwin, 1975) which was the basis for *Origin of Species*, Darwin noted that just as more commodities could be produced in a given time when labourers with different and specialized skills were used, so in "in any country, a far greater number of individuals descended from the same parents can be supported, when greatly modified in different ways, in habits, constitution and structure, so as to fill as many places, as possible, in the polity of nature, than when not at all or only slightly modified." (1975: 228) This analogy was extended to the scale of the whole planet, where a greater number of different types of organisms could be supported when diversification was large than when it was not. As a result, natural selection acting upon a varying population in circumstances of environmental change resulted in what Darwin termed "descent with modification." Significantly, there was no preset limit to the amount of these modifications over long periods of time. This process of descent with modification was graphically represented by Darwin in *Origin of Species* as a tree-like structure: (1872, 1: 167)

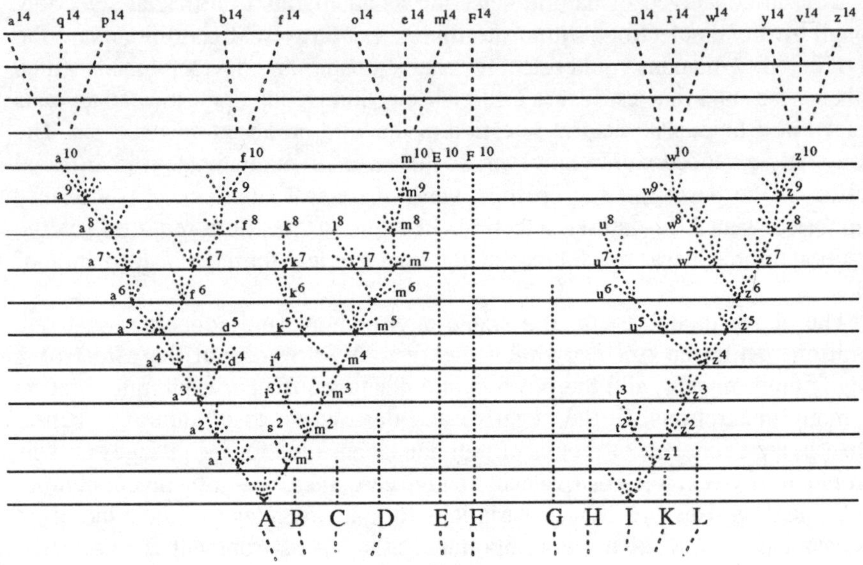

Diagram 1.1: Darwin's tree of evolution

This was most aptly described in *Natural Selection* as follows:

> The relation of all past and present beings may be loosely compared with the growth of a few gigantic trees. ... The buds and twigs may represent existing species, and all beneath their living extremities may represent extinct forms. (1975: 249)

Each level indicated by horizontal lines stood for one thousand generations, so that the whole tree represented fourteen thousand cycles of natural selection. The spreading of branches corresponded to divergence of character, and the pruning of branches to extinction of the inadapted. Divergence of character was indicated by the branches that bifurcated as at a^3 or m^2, with the most extreme varieties being preserved, while the extinction of the inadapted was indicated by branches that terminated, such as those at d^5 or k^8. The diagram represented the fact that a parent population at *A* after ten thousand generations had resulted in daughter populations a^{10}, f^{10}, and m^{10} that were descended with considerable modifications from it – enough to justify calling them distinct species. After another four thousand generations, m^{10} had diverged to the populations o^{14}, e^{14}, and m^{14}, giving rising to a new genus. The population derived from *F* was unchanged after all that time, while all the descendants of *E* had become extinct. The criteria, however, for distinguishing species and other taxa were not yet explicit, and only graphically illustrated. Darwin's analysis of taxonomic concepts will be further examined in section (c).

(b) The Continuity of the Evolutionary Process

But whatever the criteria for identifying species and higher taxa, any branch of the tree showed a continuous variation of a characteristic, illustrating the principle of the continuity of the evolutionary process, a basic postulate of Darwin's framework for evolutionary theory.[7] The principle of continuity was often referred to by Darwin as "*natura non facit saltum*"– nature does not make leaps. Darwin considered this principle to be a canon of nature and repeated the phrase on at least five occasions in the first edition of the *Origin of Species*. The following was typical of his statement of the thesis:

> It is, no doubt, extremely difficult even to conjecture by what gradations many structures have been perfected, more especially amongst broken and failing groups of organic beings; but we see so many strange gradations in nature, as is proclaimed by the canon, "Natura non facit saltum," that we ought to be extremely cautious in saying that an organ or instinct, or any whole being, could not have been arrived at its present state by many graduated steps. (1872, 2: 277)

The principle could be expressed both in the negative sense that nature made no leaps, so that there were no gaps in the path of evolution, and in the positive sense that nature was continuous, so that between any two types there was an intermediary one. This principle was not new. Aristotle, in *History of Animals* formulated it as follows: "Nature proceeds little by little from things lifeless to animal life in such a way that it is impossible to determine the exact line of demarcation, nor on which side thereof an intermediate form should lie." (588b4-14) Continuity was such that there existed organisms which were intermediate between plants and animals, as well as entities intermediate between the living and the non-living. Leibniz in the preface to *New Essays on Human Understanding* (1705) commented in the same vein: "Nothing is accomplished all at once, and it is one of my great maxims, and one of the most verified, that nature makes no leaps; a maxim which I called the Law of Continuity." (1981: ix) Though leaps were thereby excluded from nature, Leibniz admitted that gaps appeared to exist. But these apparent gaps existed only because humans were limited to the observation of species in their immediate environment. Intermediate forms might exist in unexplored parts of this world or on other worlds of which humans had no knowledge.

Though Darwin, Aristotle, and Leibniz agreed on the principle of continuity in nature, they differed on related issues. Aristotle identified the forms of human, animal, and plant life in three grades or types of soul, while Leibniz argued for a pre-established harmony where physical causation did not exist, and where species formed a "great chain of being." (Lovejoy, 1936) This principle of the continuity of nature, which for Aristotle and Leibniz was a static principle implying that nature was a plenum of species, became for Darwin a dynamic principle presupposing continuous and infinitesimal changes over time. Apparent gaps in the fossil record were produced because intermediary species had not been fossilized. Darwin believed that the saltationist alternative was unacceptable, because leaps in the course of evolution itself – and not just gaps in its record, which he of course admitted – seemed to be a concession to creationist views, where divine fiat could produce the discontinuities *ex nihilo*. Darwin's belief in continuity and the implications of this belief played a major role in his works on plants and worms.[8]

Rather than being an avocational sidelight, Darwin's books on plants were part and parcel of his evolutionary thinking, particularly his notion of the continuity of life. He was especially interested in those features of plants which they had in common with animals, such as sexual reproduction, the power of movement, and in some, the ability to capture and eat insects. He concluded his *Power of Movement of Plants* (1880) with a section concerning the radicles of plants, which he considered to play the key role in movement: "It is hardly an exaggeration to say that the tip of the radicle thus endowed, and having the power of directing the movements of the adjoining parts, acts like the brain of one of the lower animals; the brain being seated within the anterior end of the body, receiving impressions from the sense-organs, and directing the several movements." (1874: 573) This was not to say that Darwin believed that plants had brains, but given the principle of continuity, and the fact that brains had appeared in evolution among the animals, he concluded that structures similar to and foreshadowing brains must have existed among the plants, particularly those displaying the property of movement in common with animals.

In the concluding chapter of his last work, *The Formation of Vegetable Mold by the Action of Worms* (1881), the result of four decades of research, Darwin noted that though almost totally blind, completely deaf, and hardly able to smell, earthworms were nevertheless responsible for the production of the whole of the arable topsoil. But despite their lack of many sensory faculties, they "apparently exhibit some degree of intelligence instead of a mere blind instinctive impulse, in their manner of plugging up the mouths of their burrows." (1881: 374) In Darwin's view, there was a continuous line of development from plants and especially animals to humans, so that what existed in humans, such as mind and intelligence, must also exist, if only in a less developed form, in animals and plants as well. As a result, Darwin saw mind at some degree of development at every stage of the evolutionary process, a position which may be termed a "bio-psychist" point of view – the notion that some sort of mind or intelligence exists in all living things, becoming progressively more developed as evolution advanced.[9]

Darwin's biopsychism found its logical extension in Ernst Haeckel's (1834-1919) panpsychism, the theory that all entities, whether living or not, were endowed with some form of psychic qualities. Though Haeckel did not attribute consciousness in the human sense to atoms, in *The Riddle of the Universe* (1899) he did grant them "the elementary psychic qualities of sensation and will." (1899: 179) These mental qualities were unconscious ones in the case of inanimate things and conscious ones in higher organisms. The attribution of basic psychic qualities to the inanimate meant that no breach in continuity occurred with the appearance of life, and though this was a step that Darwin did not himself make, it was nevertheless consistent with his philosophical framework.

Darwin confined his analysis of continuity to the realm of the biological, but even in this restricted domain his use of the concept was not without its difficulties. In mathematics, continuity of a curve implies that there is an infinite number of points, since between any two there is a third, and so on *ad infinitum.* Darwin nowhere stated that he believed there to be an infinite number of organisms, either in the present or in the past, so that his concept of continuity cannot be understood in this sense. Darwin's concept of continuity invoked the notion of differences which were sufficiently small as to be negligible for all practical purposes. Referring to his tree diagram of evolution, reproduced on page eight, Darwin supposed that "the amount of change between each horizontal line in our diagram to be excessively small," and that only the diagram as a whole represented a "considerable amount of change." (1872, 1: 169) If the differences between individuals in a species, or between species in a genus, were sufficiently small then for all practical purposes there would be a continuous transition from one individual to the next, and from one species to the next.

For this to be possible, the differences must be quantitative so as to be measurable, for were there to be the introduction of new qualities there would be a discontinuous leap between distinct and incommensurable characteristics. Darwin's concept of the continuity of the evolutionary process hinged on the further thesis of his philosophical framework for evolutionary theory that the character of change in nature was quantitative.

(c) Nature of Change: Quantity and Quality

Quantitative change is measurable change within a given characteristic, while qualitative change is a change in the type or kind of characteristic. Historically, the distinction between quantitative and qualitative change was linked to that between primary and secondary properties. Descartes, Galileo, and Locke held that primary properties were essential to matter, independent of the mind of the observer, and measurable through experiment. They included the size, shape, and weight of the corpuscles or atoms of which all bodies were composed. Secondary properties were the perceptual characteristics accessible by sight, taste, touch, smell, and hearing, which depended on the interaction between the primary properties of the object and the sensory system of the subject. In the absence of the subject, the secondary properties or perceptual qualities would not exist. In the Newtonian worldview, everything that was scientifically worthwhile could be reduced to quantitative relationships among the small number of primary properties. This was the research program of the new science that displaced the qualitative theory of Aristotle which had held sway for almost two millennia.[10]

In accord with this general background, Darwin's theory of natural selection presupposed a theory of change which was basically quantitative in character. A change in kind implied a leap, whereas change in quantity was consistent with continuity. And as continuity was the chief distinguishing mark between Darwin on the one hand, and the saltationists and creationists on the other, he opted for quantitative change as the mode of evolutionary progress. As a result, Darwin stated in *Descent of Man* that though "the difference between the mind of the lowest man and that of the highest ape is immense," nonetheless, "the difference in mind between man and the higher animals, great as it is, certainly is one of degree and not of kind." (1871, 1: 170) This view of change was universalized in Darwin's general theory of species formation.

Small changes in the average characteristics of populations resulted in the formation of varieties, which when further modified, were termed "well marked varieties," "species in the process of formation," or "incipient species." (1872, 1:158-159) A greater amount of change resulted in a population which could be considered as a new or transformed species, as can be seen in Darwin's diagram of the tree of life reproduced on page eight. Referring to the tree diagram, Darwin noted:

> Only those variations which are in some way profitable will be preserved or naturally selected. And here the importance of the principle of benefit derived from divergence of character comes in; for this will generally lead to the most different or divergent variations (represented by the outer dotted lines) being preserved and accumulated by natural selection. (1872, 1: 88-89)

Darwin explicitly indicated that the vertical axis represented time, with each horizontal line standing for one thousand generations. Though he did not annotate the horizontal axis, he stated the following:

> Let A to L represent the species of a genus large in its own country; these species are supposed to resemble each other in unequal degrees, as is so generally the case in nature, and is represented in the diagram by the letters standing at unequal distances. ... The branching and diverging dotted lines of unequal lengths proceeding from (A) may represent the varying offspring. The variations are supposed to be extremely slight, but of the most diversified nature; they are not supposed all to appear simultaneously, but often after long intervals of time; nor are they all supposed to endure for equal periods. (1872, 1: 165)

The key phrases are "unequal degrees," "unequal distances," "unequal lengths," and "diversified nature." The "unequal degrees" of resemblance of species in a genus presupposed that they shared common characteristics which could be measured and compared quantitatively. This interpretation is reinforced by Darwin's placing the points standing for the species *A* to *L* at "unequal distances" on the horizontal axis, corresponding to the degree of difference between them in reality. Lineages of offspring were represented by dotted lines of "unequal lengths," with variations identified as "extremely slight." This too presupposed a quantitative concept of the variation involved. In the last sentence of the quotation, Darwin referred to the variations being of the most "diversified nature," so that the populations which were represented as points on distinct branches, such as a^4 and m^4, differed from *A* in modifications of different characteristics. Once again, the logic of the argument implied that the differences must be quantitative in order that the populations could be compared in reality and represented on the graph at specific points.

The horizontal axis, therefore, represented the average amount of a characteristic possessed by daughter populations descended from the ancestral stock such as *A*. The distance from *A* of a point representing a daughter population, as measured on the horizontal axis, represented the degree of divergence between the two. The particular characteristic being measured differed from lineage to lineage in reality, and from branch to branch in the diagram. Continuous variation in a single characteristic was represented by a branch of the tree, so that the whole graph gave a simplified representation in two dimensions of what should be a multi-dimensional state space of varying characteristics. The change involved in descent with modification was quantitative: increments and decrements of characteristics with the succeeding generations.[11] Taxonomic distinctions – in the present case, that of well-marked varieties which were represented in the tree diagram by a superscripted letter at one of the horizontal lines – entered only through the classifying activity of biologists:

> When a dotted line reaches one of the horizontal lines, and is there marked by a small numbered letter, a sufficient amount of variation is supposed to have been accumulated to form it into a fairly well-marked variety, such as would be thought worthy of record in a systematic work. (1872, 1: 88-89)

Classification, which made qualitative distinctions in distinguishing among different varieties, species, genera, and higher taxa, was a conceptual activity where conventions not directly provided by nature were adopted. What nature provided were individuals and their characteristics, including individual differences which could be observed and measured. Taxonomists established the criteria for the rest – varieties, species, genera, families, and further taxa.[12] These qualitative distinctions, though based on real distinctions in nature, were nonetheless the product of human conceptualization and the desire for classification:

> From these remarks it will be seen that I look at the term species as one arbitrarily given, for the sake of convenience, to a set of individuals closely resembling each other, and that it does not essentially differ from the term variety, which is given to less distinct and more fluctuating forms. The term variety, again, in comparison with mere

individual differences, is also applied arbitrarily, for convenience' sake. (1872, 1: 88-89)

Darwin's use of the phrases "arbitrarily given" and "for convenience' sake" were meant to be taken seriously. The system of classification was underdetermined by the facts of phylogeny. Descent with modification placed a constraint on the possible taxonomic schemas, though it did not uniquely determine which one was correct.[13] On this interpretation of Darwin, the evolution of species was not the same as descent with modification, though based on it. Descent with modification was the real process whereby later populations were modified with respect to an original stock, a process of quantitative change over historical time. In biology, the science that studied this process, these quantitative changes were combined with qualitative considerations, in particular, the designation of taxa introduced by the biologist's classifying activity. The result was the concept of the evolution of species – a hybrid notion which was neither an exact transcription of the actual filiation of populations, nor an arbitrary fiction divorced from real phylogenies.

Darwin did not use the term "evolution of species" in the first edition of his work, though the book ended with the word "evolved." He introduced it in a later edition, borrowing the term "evolution" from Spencer. (Bowler, 1967) "Evolution', which originally meant the ontogenetic development of individual organisms, became synonymous with "descent with modification." But when the term "evolution" – referring to a real process – is conjoined with the term "species" – referring to a conceptual distinction – in the concept of "evolution of species," it is important to respect Darwin's view on the "arbitrary" and "conventional" character of taxonomic designations, that of species in particular.

Darwin was not a realist on the question of the status of qualities as he did not admit qualitative change into the process of descent with modification. Nor was he an anti-realist with respect to qualities since he did not deny a role to qualities in his theory of the evolution of species. Qualities had a place – not in nature as it existed independently of humans – but in the classificatory schemata of the taxonomist. His position was an intermediate one, which may be termed "conceptualist," just as conceptualism was a compromise between realism and nominalism in the medieval debate over universals. Consistent with the Galilean-Newtonian background to the scientific revolution, quantitative change of primary properties was the basic process in nature, while qualitative considerations played a role similar to that of secondary properties – products of the human mind in interaction with the primary properties of the external world. The emergentism of

Lloyd Morgan will attempt to provide a more realistic basis for novel qualities – those of life and mind in particular – in nature itself.

(d) Source of Change: Naturalism or Supernaturalism

A third component of Darwin's framework for evolutionary theory was the principle that the source of change was natural, and on this essential point Darwin differed markedly from his predecessors in biology. The basic question was whether biological change had its origin within nature itself, or whether the source of that change was supernatural, in Deity or God. The English natural theology tradition asserted that God was the creator and supervisor of Nature, and therefore the originator of species. God's goodness and power were manifested in nature, and an examination of species showed that they were designed by a kind and wise creator to flourish in their given environments. A typical and influential expression of this viewpoint was John Ray (1627-1705) in *Wisdom of God as Manifested in the Works of Creation* (1691). By the end of the 18th century, natural theology became the dominant philosophical-theological framework for science, both biological and natural, and William Paley (1743-1805) in *Natural Theology* (1807) consolidated this trend in a widely read popularization of the theory.[14] The *Bridgewater Treatises* appeared just as Darwin was completing his voyage of discovery around the world, and marked the high point, and paradoxically the beginning of the end of the domination of natural theology as a philosophical framework for biology.[15]

It was against this tradition that Darwin rebelled, though he was not the first to do so. In the preceding century, David Hume (1711-1776) had turned his skepticism to the question of natural theology in *Dialogues Concerning Natural Religion* (1779). Hume argued that from multiple effects, a single cause could not be adduced, nor did a perfect creator follow from the existence of imperfect creations. When Darwin argued against a natural theological foundation for biology, and replaced it with one based on natural selection and a non-theistic framework, he did not break new ground in philosophy, but he did launch the final attack on natural theology and special creation in the realm of biology.

As a young undergraduate at the University of Edinburgh, Darwin was in contact with ethical theories and philosophies of mind which tended towards naturalism. (Manier, 1978) A major influence on Darwin upon his return to England after his sea voyage on the H. M. S. Beagle was the French philosopher Auguste Comte (1798-1857), who argued in his *Positive Philosophy* (1830-5) that human thought went through three stages: the first and most primitive was the theological, where divine beings were considered as the source of all events; the second was the metaphysical, where the accent was on the nature or essence of things; the third and highest was the positive, which dropped all reference to divine causes as well as to natures and essences, in favour of constant correlations of antecedent and subsequent events. Comte provided a methodological basis for a naturalistic world view in science, even if in biology he was not an evolutionist. Darwin read and commented on an English review of Comte's work during the time he was working on his evolutionary notebooks.[16] (Barrett, 1974) The influence of Comte can be seen in the *M Notebook* as follows:

> Now it is not a little remarkable that the fixed laws of nature should be / universally / thought to be the will of a superior being, whose natures can only be rudely traced out. When one sees / this, one suspects that our will may / arise from / as fixed laws of organization. – M. le Comte argues against all contrivance – it is what my views tend to.[17] (M.69 in Barrett, 1974: 18)

The following quote, though only programmatic, also indicates the profound influence of Comte on Darwin's whole view of biology: "M. le Comte's idea of theological state of science, grand *idea*: ... as soon as any enquiry commenced, for instance probably such a thing as thunder would be placed to the will of God. – Zoology itself is now purely theological." (N.12 in Barrett, 1974: 72) Darwin was clearly dissatisfied with this theological state of zoology. In the notes to Maculloch's *Proofs and Illustrations of the Attributes of God* (1832) which have been transcribed by Paul Barrett under the title "Essay on Theology and Natural Selection," Darwin rejected any theological explanation of the cause of species: "The explanation of types of structure in classes – as resulting from the will of the deity, to create animals on certain plans, – is no explanation – *it has not the character of a physical law* /& is therefore utterly useless. – it foretells nothing / because we know nothing of the will of the Deity, how it acts & whether constant or inconstant like that of man. – the cause given we know not the effect." (in Barrett 1974: 157-158) The rejection of supernatural agency and theological explanation was a constant feature of Darwin's further writings. Darwin meant to transform zoology from its theological to its positive stage, a task he was to accomplish with his theory of evolution by natural selection.

The influence of Comtian positivism on Darwin may also account for his attitude towards Lamarck. Lamarck's theory of evolution was set out in *Zoological Philosophy* which appeared in 1809. There, Lamarck argued for three factors of evolution: firstly, on-going spontaneous generation that brought about living organisms from inanimate matter; secondly, a perfecting principle according to which organisms become more complex, especially as concerns their internal organs, the heart and the brain in particular; and thirdly, the inheritance of acquired characteristics to provide for an automatic adaptation of organisms to modified circumstances. This accounted for local deviations from the expected result of the perfecting principle acting alone, and explained why no species became extinct. Darwin was familiar with Lamarck's theory both before and during his voyage on the H.M.S. Beagle. One of his teachers at Edinburgh, Edward Grant, was a supporter of Lamarck and discussed his theories with Darwin. While on his voyage of discovery, Darwin received the third volume of Lyell's *Principles of Geology* where Lyell carefully though critically set out Lamarck's system. Yet Darwin in his *Autobiography* , written in 1876, indicated that Lamarck's theory had little specific effect on him, though it may have disposed him more favourably to the general notion of evolution. Speaking of an occasion when Grant discussed Lamarckism with him, Darwin stated:

> I listened in silent astonishment, and as far as I can judge without any effect on my mind. I had previously read the *Zoonomania* of my grandfather [Erasmus Darwin], in which similar views are maintained, but without producing any effect on me. Nevertheless, it is probable that the hearing rather early in life such views maintained may have favoured my upholding them under a different guise in my *Origin of Species*. (1974: 26)

This can be understood in terms of Comte's theory of stages. In Lamarck's theory it was of the "essence" of species to progress towards perfection and so undergo transmutation; no physical mechanism was provided. Darwin rejected such an approach as metaphysical, while considering that his own work provided a positive approach based on natural selection. Influenced by Comte in this respect, Darwin would have no more use for a metaphysical principle of species perfection than for a theological one of special creation, and he would evaluate his own theory as one which, though sharing a general belief in species transmutation with Lamarck, presented the argument in a "different guise," one which was positive rather than metaphysical.

A reinforcing influence on Darwin's naturalism was furnished by Lyell's principle of uniformity. Lyell's thesis, according to which past causes were similar to those presently acting, was essential in his transformation of geology into a wholly natural science, and was interpreted by Darwin in a sense more radical than that of his mentor. In biology Lyell continued to accept divine intervention in the creation of species, and believed that variation in organisms always respected definite and small limits. Darwin, however, applied the principle of uniformity to biology and the origin of species as well. He was completely uniformitarian, more Lyellian than Lyell himself. The net result of the combined influence of Comte and Lyell was the reinforcement of Darwin's naturalism.

(e) Materialism and the Mind-Brain Relation

It may seem a small step from naturalism to materialism. The existence of an immortal and immaterial mind presupposes a supernatural and divine source, and since naturalists reject supernaturalism, it would seem only reasonable for them to conclude that mind must be both mortal and material. But the step was one that many naturalistic philosopher-scientists, Aristotle first and foremost, did not take. In Darwin's time and the century preceding, materialism was linked to atheism, sedition, and immorality – religious, political, and ethical charges of extreme gravity. In his will establishing the lecture series on natural theology that bore his name, Robert Boyle had specifically instructed the lecturers to combat these evils. In philosophy Hobbes, a materialist, was to be refuted, not followed. Though Locke could ask whether God might not superadd the quality of thought to otherwise brute, inert matter, this was not a thesis of his metaphysical system, but merely a speculation. Berkeley, with more assurance in his own convictions, argued for an idealistic view. Everything was an idea in some mind, either finite and human, or infinite and divine. Turning the tables on materialism, he boldly stated that the brain, like any other thing, was an idea in the mind. Hume, with his characteristic skepticism, cast doubt on the significance of the very concept of substance, whether material or mental, and restricted inquiry to phenomena available to the senses. Joseph Priestley (1733-1804), in his *Disquisitions Relating to Matter and Spirit* (1776) was alone among prominent English scientists in the 18th Century in defending an explicitly materialist thesis that mind was a function of the brain.[18] Priestley held that the whole brain had a property – the ability to think – that none of its parts possessed, anticipating a major thesis of emergentism. He did not, however, analyze this point any further, though he repeated it on numerous occasions in his *Disquisitions* and other works.[19]

In France, however, there was a school of materialists with La Mettrie in *Man a Machine* (1747) an early representative, and Lamarck a later one.[20] These materialists adopted a medical metaphor, claiming that the brain secreted thought just as the liver secreted bile, and it was precisely this image that Darwin would himself use in his notebooks and other personal writings of the 1836-38 period, as he set out what he considered to be a materialist theory of mind. In his marginal notes to John Abercrombie's *Inquiries* (1839), Darwin said: "By materialism I mean merely the intimate connection of kind of thought with form of brain – like kind of attraction with nature of element." (in Manier, 1978: 223-224) In his *M Notebook*, Darwin referred to materialism on a number of occasions: "It is an argument for materialism that cold water brings on suddenly in head, a frame of mind analogous to those of feelings which may be considered as truly spiritual." (M:19, in Barrett, 1974: 9) In this case, Darwin interpreted the fact that a physical cause brought about a mental effect as indicating that the mental effect was itself physical in nature. Again: "To avoid saying how far I believe in Materialism, say only that emotions, instincts, degrees of talent, which are hereditary are so because brain of child resembles parent stock. – (and phrenologists state that brain alters)." (M:57, Barrett, 1974: 16) The theme of the heritability of mental states because of the resemblance of the child's brain to its parents' was take up again in the *C Notebook*, where Darwin proclaimed, chastising himself for so radical a view: "Thought (or desires more properly) being hereditary it is difficult to imagine it anything but structure of brain hereditary, analogy points to this – love of the deity effect of organization, oh you materialist." (C:166, Barret, 1974: 190).[21]

Darwin considered as materialist the thesis that mental capacities were a product of the organization of the brain. This theory was consistent with the mechanistic outlook of much of 19th century science, which adopted a theory of cause-effect correlations. Given a structure which acted as a cause – in this case the brain – there followed a corresponding function as effect – in this case mental states. The product was similar in kind to the producing mechanism, and as the brain was material, so was thought. Darwin's materialism could not be more evident, though during his lifetime it remained private, limited to the intimacy of his notebooks. In his public statements on the mind, such as *The Expression of Emotions in Man and Animals*, which appeared in 1872, there was nothing to contradict a materialist point of view, though no explicit expression of it either.

(f) The Scope of Evolution: Man, Mind, and Morals

Darwin's theory of evolution radically modified the foundations of biology, but the question was immediately posed whether the theory could be applied in other domains as well. This was the problem of the scope of evolution. As the domain of application of the concept of evolution was broadened, it became more and more a philosophical theory of change, not only applicable to life, but to matter, mind, and society as well. Herbert Spencer, in his multi-volume *Synthetic Philosophy* (1855-93) set out a philosophical theory of evolution which was applied to all domains of reality, including the inorganic, the organic, and the human. But Darwin was more cautious and slower to jump into the fray, even concerning the application of evolution by natural selection to man, mind, and morals, with "man" in its 19th century context as a term for *homo sapiens*.[22] In the first edition of *Origin of Species*, Darwin noted that "[m]uch light will be thrown on the origin of man and his history," (1872, 2: 314) though as of 1838, some 20 years earlier, he had already become convinced that humans and the great apes shared a common progenitor. It was in *Descent of Man and Selection in Relation to Sex* (1871), a dozen years after the publication of *Origin of Species*, that this origin was presented. Following the publication of Lyell's *The Geological Evidences of the Antiquity of Man* in 1861 and Huxley's *Man's Place in Nature* in 1863, the *Descent of Man* established evolution in the field of human studies as well. But the extension of the evolutionary concept to the biological origin of humanity was a mere application of the theory, making explicit what was already implicit in the *Origin of Species*. *Homo sapiens* as a species was, like any other species, subject to the biological law of evolution.

Darwin also extended the scope of evolution to the mind, and in *The Expression of Emotions in Man and Animals* he laid the basis for comparative psychology, studying in detail the wide range of sentiments common to humans and animals. Darwin collaborated with George John Romanes on the comparative study of instinct, and Darwin's essay "On Instinct" was published posthumously in Romanes' *Mental Evolution in Animals* (1883). The extension of evolutionary theory to man and mind was relatively straightforward, at least in principle, though details to this day remain to be worked out.

More controversial and problematic was Darwin's attempt to extend the scope of his theory to the evolution of morals. In *Descent of Man* he presented his case for the existence of social instincts which aimed at the common good. "Man" was a social animal, inheriting from his animal past "a tendency to be faithful to his comrades, and obedient to the leader of his tribe," as well as a tendency to "defend, in concert with others, his fellow men," and "to aid them in any way which did not too greatly interfere with his own welfare or his own strong desires." (1871: 149) Combined with "instinctive sympathy" which caused humans to seek the approbation and avoid the disapprobation of their peers, this led to the development of moral values in the service of the general good. The "general good" was defined as "the rearing of the greatest number of individuals in full vigor and health, with all their faculties perfect, under the conditions to which they are subjected," (1871: 149) and was distinguished in this biological context from the related but distinct notion of the general happiness of the species, which involved further psychological considerations.

Darwin initially presented a humanistic and optimistic view of moral evolution. Humans would value the "dignity of humanity" above individual selfish desires, and the higher, social instincts, with their derived virtues, would win out in the struggle with the lower, individual impulses. Darwin concluded: "Looking to future generations, there is no cause to fear that the social instincts will grow weaker, and we may expect that virtuous habits will grow stronger, becoming perhaps fixed by inheritance. In this case the struggle between our higher and lower impulses will be less severe, and virtue will be triumphant." (1871: 169) This conclusion was then brought in line with Darwin's theory of natural selection. Although a high standard of morality might give no benefit to a particular member of a tribe, and indeed might be destructive of existence as in the case of sacrificing one's own life for the sake of another's, the matter was different when examined from the point of view not of the individual, but of the collectivity:

> A tribe including many members who, from possessing in a high degree the spirit of patriotism, fidelity, obedience, courage, and sympathy, were always ready to aid one another, and to sacrifice themselves for the common good, would be victorious over most other tribes; and this would be natural selection. At all times throughout the world tribes have supplanted other tribes; and as morality is one important element in their success, the standard of morality and the number of well-endowed men will thus everywhere tend to rise and increase. (1871: 179)

A decidedly less optimistic note was sounded when Darwin discussed the application of natural selection in civilized countries. The context of Darwin's discussion had now shifted from that of the evolution of solidarity among early humans, to that of modern society and its welfare policies. Darwin cited a series of measures that were being taken to protect "the imbecile, the maimed, and the sick," so that the inadapted, as the result of social protection, were now being preserved and allowed to propagate, rather than being eliminated through natural selection. Darwin was struck by the paradox that as the result of conscription and war, "healthy and vigorous" young men were placed in circumstances of corruption and death, while the feebler remained at home to marry and reproduce. Darwin worried that the inevitable result of such a subverting of natural selection would be the impoverishment of the population stock, and a decline in the quality of civilization.

Two strains of thought in Darwin's work were now in conflict. On the one hand, when he discussed early hominid evolution, Darwin had optimistically argued for the triumph of social values over individual desires. Groups practicing social solidarity would be selected over groups unwilling or unable to do so. On the other hand, when he discussed modern society, Darwin became pessimistic, fearing that welfare measures would subvert natural selection and lead to a decline in standards. In the first case, Darwin was at the verge of admitting a second factor of evolution, a cooperative group factor. In the second case, Darwin was merely applying the individual factor of natural selection to a group setting.

This indecision – whether to admit a further factor of cooperation at the social level, or merely extend the existing factor of competition which he had found so useful at the biological level – was to have lasting impact. Those who applied the struggle for existence to the ethical level, and considered it as the motor of social development, moved to what became known, somewhat inappropriately, as a Social Darwinist position, countenancing cut-throat competition and rejecting social welfare. (Montagu, 1952) Applied to countries, this provided a theoretical justification for national chauvinism and war, though it is likely that Darwin would have rejected these extreme consequences. Those who took the route of adding a new factor of social cooperation developed an ethics and social philosophy based on harmony and sympathy. (Drummond, 1887 and Kropotkin, 1902)

Darwin's view of the framework for evolutionary theory thus included the following elements: the principle of continuity, according to which there were no gaps, leaps, or jumps in the course of evolution; a quantitative view of the character of change, relegating qualitative distinctions to the mind of the biologist; a naturalistic view of the source of evolution, denying any supernatural intervention in the evolutionary process; a materialist ontology, based on a mechanistic view of the mind-brain relationship; and the extension of the scope of evolution to include at the very least man, mind, and morals. Darwin's collaborators and colleagues – Spencer, Huxley, Wallace, and Romanes – had differences with him not only concerning the scientific question of the factors of evolution, but also concerning the philosophical framework for evolutionary theory.

CHAPTER TWO

Herbert Spencer: Philosophy of Evolution

Herbert Spencer (1820-1903) – unlike Darwin, Huxley, Wallace, and Romanes – was not a scientist. Though he wrote *Synthetic Philosophy* (1855-1893, in 10 volumes), one of the epic multi-volume treatises on philosophy in the Victorian period – the epic if length is taken into account – he was not trained as a philosopher either.[1] He was in fact an engineer by formation and a publicist by trade, living from the proceeds of his writings. He became one of the most important of the Victorian philosophers of evolution and his work commanded attention and interest in America as well as in England.[2] Today Spencer is best remembered as one of the founders of sociology and a defender of Social Darwinism, though it will be argued that this latter charge is ill-founded. Over-rated as a philosopher during the latter part of the 19th century, Spencer's reputation has declined precipitously with the 20th, and is arguably as under-estimated today as it was over-estimated in his own time.

Spencer was in communication with Darwin and Huxley, and his comments on biological subjects often appeared in the major monthly reviews and in the letters column of *Nature*. One of Spencer's important contributions to evolutionary theory was the term "survival of the fittest', which Darwin also adopted in the *Origin of Species*.[3] But more than in terminology, Spencer innovated in philosophy. He first broached his philosophical concept of evolution in "The Development Hypothesis" which he wrote in 1852 – six years before the joint presentation of their work on evolution by Darwin and Wallace to the Linnean society, and seven years before the *Origin of Species* provided an extended discussion of the scientific evidence for evolution.[4] In this respect, Spencer was one of the pioneers of evolution in England, though not the first, having been preceded in this by Robert Chambers (1802-1871) in his anonymous *Vestiges of the Natural History of Creation* (1844).[5] In the "The Development Hypothesis" (1852), Spencer argued, like Lamarck before him, that it was the modifying influence of the environment which brought about species change. He asserted that the proponents of the development hypothesis, or evolutionary theory in subsequent terminology, could claim the following:

> They can show that any existing species – animal or vegetable – when placed under conditions different from its previous one, immediately begins to undergo certain changes fitting it for the new conditions. They can show that in successive generations these changes continue;

> until ultimately the new conditions become the natural ones. They can also show that in cultivated plants, in domesticated animals and in the several races of man, such alterations have taken place. (1852: 3)

This modifying influence of the environment could produce such enormous differences between daughter populations and the parental stock that "it is a matter of dispute whether some of these modified forms are varieties or separate species." (1852: 3) Extrapolated over a long period of time and under greatly modified environments, such a process could result in very great changes: "And thus they can show that throughout all organic nature there is at work a modifying influence of the kind they assign as the cause of these specific differences: an influence which though slow in its action, does, in time, if the circumstances demand it, produce marked changes – an influence which, to all appearance, would produce in the millions of years and under the great varieties of condition which geological records imply, any amount of change." (1852: 4) This statement was surprisingly Darwinian concerning the rate of change, with small modifications accumulating over long periods of time to bring about "any amount of change," though it was Lamarckian in its identification of the cause of that change as the direct action of the environment.

In defending evolution, Spencer argued that it was simpler, and therefore more rational, to explain millions of varieties and species through one developmental principle acting as a law of nature, rather than through millions of separate, special acts of creation. A version of Occam's razor was at work: one natural principle was preferable to many supernatural actions. Spencer also deployed arguments from analogy, one from geometry and the other from embryology. Just as there was a mathematical transformation of the conic sections – circle, ellipse, parabola, and hyperbola – one into the other, so too there was one biological process, that of "development," which produced the progressive transformation of one organic species into the next. In embryology, just as a seed produced a tree and a single cell produced a human, "there is nothing absurd in the hypothesis that under certain other influences, a cell may, in the course of millions of years, give origin to the human race." (1852: 6) Though the principle of simplicity of explanation and the analogies from mathematics and embryology were arguments for Spencer's theory, he as yet had no universal mechanism upon which to found his belief in a developmental process in all domains of nature. This was the weakness of his position in comparison with that, based on the factor of natural selection, which Darwin had already formulated for the biological realm, but not yet published.

(a) Transformation of the Homogeneous into the Heterogeneous

By 1857, in "Progress: Its Law and Cause," Spencer used the terms "progress" and "evolution" to replace "development." His identification of evolution and progress was typically Victorian, reflecting the unstated assumption that humanity had advanced from its origin in barbarism to its apex in English middle class society. Spencer's point of departure for his analysis of evolution was embryological, but rather than a mere analogy as in 1852, he could now point to a more concrete principle, that of von Baer's transformation of the homogeneous into the heterogeneous:

> The investigations of Wolff, Goethe and von Baer, have established the truth that the series of changes gone through during the development of a seed into a tree, or an ovum into an animal, constitute an advance from homogeneity of structure to heterogeneity of structure. ... It is settled beyond dispute that organic progress consists in a change from the homogeneous to the heterogeneous. (1857: 9-10)

The basis of these changes from homogeneous to heterogeneous was the constant differentiation of the cell, a basic physiological mechanism in embryology. Spencer's move was to generalize this transformation of the homogeneous into the heterogeneous to a universal mechanism touching all domains of existence – "[I]n the development of the Earth, in the development of Life upon its surface, in the development of Society, of Government, of Manufacture, of Commerce, of Language, Literature, Science, Art, this same evolution of the simple into the complex, through successive differentiations, holds throughout." (1857: 10) Differentiation was then followed by the integration of the parts of the newly differentiated system, and the cycle began once again. Spencer's system was a full fledged philosophy of evolution, with a metaphysical mechanism and a potentially universal domain of application. Nonetheless, Spencer left out astronomical and cosmological development in his list, though he did at one point touch on the Kant-Laplace nebular hypothesis for the formation of the solar system. Spencer had little interest in physics and astrophysics, and references to these domains are rare in his work, despite the fact that he admitted the inorganic as the domain preceding the organic. Though Auguste Comte in his *Positive Philosophy* began his list of the factual sciences with the astronomical, Spencer, who in many ways followed Comte – though he vociferously denied this – began his system in practice, if not in theory, with the biological.

Spencer recognized that von Baer's principle, now generalized beyond its specific domain of application in embryology, had itself to be explained. A first level of explanation was found in the principle of multiple effects: "Every active force produces more than one change – every cause produces more than one effect." (1857: 38) Assuming that every active force produced more than one effect, there was a consequent "ever growing complication of things." (1857: 38) This still left unanswered a further question: Why did a given cause have more than one effect? If this principle could not be explained, then its role in providing an adequate grounding for the transformation of the heterogeneous into the homogeneous had not been achieved.

Spencer did not answer this question, but instead he shifted his ground to another principle, that of unstable equilibrium, and argued that the source of change lay in the fact that the equilibrium of any entity was necessarily disturbed by external forces, so that homogeneity was a temporary situation of dynamic equilibrium. As a result of this action of external forces on the unstable equilibrium of the homogeneous, two results were possible: either the thing would be broken down into unrelated parts as its internal cohesion was overcome, resulting in chaotic heterogeneity, or changes would be induced that did not destroy, but only modified the organization of the components, leading to "those first differentiations which constitute incipient organization." (1857: 83) There was an uneasy alliance of the two principles of multiple effects and the instability of the homogeneous which later presentations of the theory merely restated rather than resolved.[6]

In his *First Principles* (1862), Spencer accepted the traditional 19th century view that natural science had as its subject matter the study of matter in motion. Based on analogies from the study of the solid, fluid, and gaseous states of matter, he concluded that the integration of matter was consequent upon the dissipation of motion. His reasoning followed from the fact that an increase in the motion of component parts resulted in disintegration, as in the transition from solid to liquid, or from liquid to gas. Integration, as the opposite of disintegration, therefore involved the decrease or dissipation of motion. This motivated his definition of evolution: "Evolution is definable as a change from an incoherent homogeneity to a coherent heterogeneity, accompanying the dissipation of motion and the integration of matter." (1862: 359)

Spencer also recognized that the process of change could go in the reverse sense as well. This process, which he called "dissolution," was the complement to evolution, featuring disintegration rather than integration: "Evolution under its simplest and most general aspect is the integration of matter and the concomitant dissipation of motion; while Dissolution is the absorption of motion and concomitant disintegration of matter." (1862: 288) The dual concepts of integration and disintegration also allowed Spencer to distinguish three stages of change in living things, depending on whether the organism was dominated by integration, disintegration, or a combination of the two. In the first stage of development, that of the youth of an organism, integration predominated in the form of growth. In the second stage, that of maturity, there was an alternation in the dominance of integration and disintegration. In the third stage, disintegration dominated, leading to senescence and the death of the organism. It was characteristic of organisms as opposed to inanimate things that the two complementary processes were both present, resulting in what Spencer called compound as opposed to simple evolution. As a result, the development of organisms was more complex than that of non-living entities.

Spencer returned to the question of the mechanism of evolution in "The Factors of Organic Evolution" (1886), where he situated himself with respect to Darwin and Lamarck. The basic factor of evolution was the transformation of the homogeneous into the heterogeneous, generalized from von Baer's embryological principle. The secondary factors were Darwinian natural selection and Lamarckian use-inheritance. The basic factor was responsible for the origin of living matter or protoplasm, new species, and mind itself, as less complex entities were differentiated into more complex ones. Natural selection come into effect only once life had appeared, to select the fittest for survival in the struggle for existence. As life became more complex in the course of human psychological and social development, the Darwinian factor of natural selection then took second place to Lamarckian use-inheritance, through which favourable mental and ethical characters were preserved. The transformation of the homogeneous into the heterogeneous was the source of the inorganic, the organic, and the human, while Darwinian natural selection played no more than a secondary role in the organic domain, and was replaced even in this subsidiary role by Lamarckian use-inheritance in the realm of the human.

(b) Inorganic, Organic, and Super-organic Evolution

Spencer's most significant contribution to evolutionary philosophy was to distinguish three domains of reality. Spencer called these domains the "inorganic," or the domain of the physico-chemical before life, the "organic," which was the biological domain and included mind, and the "super-organic," the domain of the social and the ethical. In the introduction to *The Principles of Sociology* (1885), Spencer therefore distinguished three phases of evolution, indicating in passing that his system, massive as it was at 10 volumes, was missing two preliminary ones on astronomical and geological evolution:

> The first kind, Inorganic Evolution, which, had it been dealt with, would have occupied two volumes, one dealing with Astrogeny and the other with Geogeny, was passed over because it seemed undesirable to postpone the more important applications of the doctrine for the purpose of elaborating those less important applications which logically precede them. The four volumes succeeding *First Principles*, have dealt with Organic Evolution; two of them with those physical phenomena presented by living aggregates, vegetal and animal, of all classes; and the other two with those more special phenomena distinguished as psychical, which the most evolved organic aggregates display. We now enter on the remaining division – Super-organic Evolution. (1885: 3)

Organic evolution reached its apogee with the development of parental cooperation in the raising of progeny. Super-organic evolution then proceeded to social organization beyond the family unit, including the division of labour in economic activity, the establishment of social and political institutions, and the development of ethics to guide human conduct. Like Darwin, Spencer invoked continuity and gradualism in his discussion of the transitions from one domain to another:

> Of course no absolute separation exists [between the Organic and the Super-organic]. If there has been Evolution, that form of it here distinguished as super-organic must have come by insensible steps out of the organic. But we may conveniently mark it off as including all those processes and products which imply the coordinated actions of many individuals. (1885: 4)

The distinction between the inorganic, the organic, and the super-organic will be taken up by Lloyd Morgan in his discussion of the three stages of emergent evolution – matter, life, and mind. Emergent evolution will sharpen this distinction by a more rigorous recognition of the qualitative novelty involved in each domain, and by introducing the theoretical notion of levels of reality, neither of which were present in Spencer's work. Spencer, having made the conceptual distinction between the three stages, then diminished their importance by invoking continuity and gradualism, so that it was in large measure a question of terminology where one ended and another began. Nonetheless, he recognized that cooperation, rather than competition, became significant in the realm of the super-organic – a thesis of importance for the development of his evolutionary ethics.

(c) Evolutionary Ethics

Spencer was an enthusiastic proponent of evolutionary ethics – in a less ambiguous sense than Darwin, and quite unlike Huxley, who rejected as contradictory the idea of combining evolution and ethics. Spencer set out his views in the first volume of his *Principles of Ethics*, also issued under the title of *Data of Ethics*.[7] Spencer has been derided in current views of 19th century thought as a Social Darwinian, proclaiming the rule of the strongest as the social equivalent of the survival of the fittest. Yet a reading of his *Data of Ethics* belies this conventional designation. To the contrary, Spencer accorded considerable importance to mutual aid and cooperation, though he rejected state supported welfare measures. He was a typical liberal free trader, befitting his lower middle class origins, and held that the highest product of evolution was the enterprising Victorian. But he should no be confused with the more radical American social Darwinians such as the American sociologist W. G. Sumner, and the German biologist Ernst Haeckel who condoned cut-throat competition and national chauvinism. (Montagu, 1952)

Spencer based his theory of evolutionary ethics on an analysis and characterization of conduct, defined as those actions which aimed at adapting means to ends. Not all actions were conduct, and Spencer excluded actions such as epileptic fits, since they served no goal of the organism. Nor was conduct necessarily conscious, and Spencer included as conduct the unconscious actions of simple organisms which aimed at the survival of the individual. He traced in very schematic terms the evolution of conduct, providing examples from the infusoria, molluscs, and fish, through "savage" tribes to civilized society. Spencer concluded that the evolution of conduct had resulted in greater success in three respects: the preservation of the individual life of the organism, the protection of progeny by parents, and the maintenance of the species. Evolution was not a purely egoistic concern, since it fostered the family and the species as well as the individual.

Moreover, the highest development of group behaviour was characterized by mutual aid and peaceful cooperation, a fourth aspect of the evolution of conduct reached only in some human societies. He noted that "the limit of evolution can be reached by conduct only in permanently peaceful societies." (1879: 19) Early humans – in Spencer's sociological terms, "militant societies" – were not able to fully achieve the four ends of conduct, and it was only with modern civilization or "industrial society" that they could be harmonized: "And there has followed the corollary that conduct gains ethical sanction in proportion as the activities, becoming less and less militant and more and more industrial, are such as do not necessitate mutual injury or hindrance, but consist with, and are furthered by, co-operation and mutual aid." (1879: 21) That the man who coined the phrase "survival of the fittest" also used the terms "cooperation" and "mutual aid" indicated the extent to which he considered evolution to have been modified as it achieved its height of development in "industrial societies," and contradicts the view of Spencer as an exclusive proponent of competition.

Spencer went on to define moral terms as properties of conduct, not abstract concepts which could not be further analyzed as the intuitionists believed, or supernaturally derived, as argued by those who based ethics on religion. A good act was one which fostered the evolution or development of the four aspects of conduct previously mentioned. Spencer noted that ethical actions which promoted and supported life increased both the duration or length of individual lives as well as what he termed the "quantity of life" by which he meant the variety and complexity of life as measured by the criterion of the balance of pleasure over pain. Here Spencer was in agreement with the utilitarians, though he differed from them in an important respect. For the utilitarians – Spencer mentioned Mill in particular – balance of pain over pleasure had no further basis. There was just the empirical calculation of quantities positive and negative.

But for Spencer there was a basis for the growing preponderance of pleasure over pain which he believed characterized both modern society and ethical action. This was evolution, which through the continuous transformation of the homogeneous into the heterogeneous, and the development of "permanently peaceful" industrial societies, provided the means for more complex beings to extend and enjoy life. Spencer considered the utilitarians as crude empiricists, while he considered that his own utilitarianism had a higher, rational basis. There was an analogy to the history of science which he found useful in this respect. Astronomy in its early empirical stage was concerned with data gathering and making predictions – the occurrence of eclipses and the tracing of planetary motions based on regularities in the data. But more advanced astronomy had laws of nature on which to base its predictions. It thus achieved a rational stage, as evidenced in the derivation of Kepler's elliptical planetary orbits from Newton's law of gravitation. Analogously, evolution as a developed theory provided the rational basis for the new science of ethics. While accepting the utilitarian criterion of balance of pleasure over pain, evolution provided ethics with a lawful basis comparable to that provided by Newton's law of gravitation in astronomy.

(d) Monism, Science, and Religion

Spencer, as with the other English evolutionists to be discussed, with the notable exception of Wallace, embedded his evolutionary views in a monistic ontology. His concept of substance was set out in the first book of the *Synthetic Philosophy*, entitled the "Unknowable", where Spencer's goal was to reconcile science and religion. He identified their common core, or shared source, as an unknowable mystery which was the substance underlying all things and their evolution. Spencer was obviously predisposed to accept science as a genuine body of knowledge, as his whole philosophy was one based explicitly on the sciences, those of biology, psychology, and sociology in particular. Moreover, he noted that "of all the antagonisms of belief, the oldest, the widest the most profound and the most important is between Religion and Science." (1867: 25) It was therefore not at all obvious that he would find something positive in religion.

His argument for a core of truth in religion involved the fact that religious ideas of one kind or another were "almost universal." Moreover, those tribes which had no religious ideas were at a lower level of general intellectual development than those that did, and so Spencer concluded that, having reached a certain minimal level of intellectual development, religious ideas were inevitable and universal. The universality of such beliefs and their independent evolution in different tribes could not be fortuitous; there must be some "deep-seated" reason for this. Something that has evolved over time could not be wholly wrong or useless; otherwise it would not have continued to evolve: "We may be sure, therefore, that religions, though even none of them be actually true, are yet all adumbrations of a truth." (1867: 31) This genetic argument set up the first limb of Spencer's argument – that there was some truth, more abstract than the doctrines of any one particular religion, and common to all.

In the second part of the argument similar reasoning was applied to science. Spencer believed that the basic propositions of science were not absolutely true. In science, there were different theories of matter, space, time, force, and motion. All of these, though containing a degree of truth, did not exhaust it. In considering the Newtonian view of matter, Spencer found fault with its assumption of complete elasticity, which violated the principle of continuity. Were matter perfectly elastic, instantaneous changes of velocity must occur in the collisions of bodies, violating the principle of continuity. Similarly, the Boscovitchian theory of point masses was rejected, for how could the addition of a finite number of infinitesimal point masses give a finite gross body? For Spencer, arguments of this sort pointed to an inadequacy not only in current conceptions of matter, but in any possible conception of matter. This applied as well to other basic scientific conceptions, such as time, space, motion, and force. "Ultimate Scientific Ideas," he concluded, "then are all representations of realities that cannot be comprehended." (1867: 78)

There were now two abstract truths which could not be adequately defined: the core truth of religion, and the ultimate scientific ideas. Religion viewed reality from the standpoint of the supernatural, science from that of the natural. Spencer found the reconciliation of the two in their common admission that there was a shared ground which neither could define nor make precise. This Spencer termed the "Unknowable," the "Unseen Reality," the "Ultimate Cause," and the "Creative Power." This Unknowable doubled as substance in his system and seen from the religious viewpoint it was supernatural, while seen from the scientific viewpoint it was natural: "While our consciousness of Nature under the one aspect constitutes Science, our consciousness of it under the other aspect constitutes Religion." (1867: 116)

A second major distinction was that between the subjective, the self, and consciousness on the one hand, and the objective, the not-self, and the unconscious on the other. The Unknowable, which was the common core of science and religion, manifested itself under one aspect as subject and under the other as object: "In brief, our postulates are: – an Unknown Power; the existence of knowable likenesses and differences among the manifestations of that Power; and a resulting segregation of the manifestations into those of subject and object." (1867: 163) Such an ontology was a "neutral" and "dual aspect" monism, in the terminology used later by Bertrand Russell: neutral because the substance it postulated had no definite properties, and dual aspect because the substance manifested itself as the correlated aspects of the physical and the mental.

Substance monism and aspect dualism were essential to Spencer's treatment of the mind-body relation. In *Principles of Psychology* (1870), mind was initially delineated as being composed of "feelings and the relations between feelings." (1870: 163) These feelings and their relations were correlated with the nerves and connecting fibres of the brain. Mind developed from the stage of mere sentiency or sensitivity to that of sensations and feelings, and then to thought and intelligence. As the feelings become more complex and took on more definite forms in the higher animals, intellect arose to combine and coordinate them. Feelings corresponded to the quantity of mind, intellect to its complexity of organization, and these were correlated in the nervous system respectively to the quantity of brain, and the complexity of motions in the brain.

Though he recognized the fact of this correspondence of mind and brain, Spencer held that to inquire into the ultimate nature of that correspondence was to ask a question which was unanswerable. Psychology could investigate the composition of mind as the psychic aspect of the underlying substance, just as physiology could investigate the composition of brain as the physical manifestation of the underlying substance. Moreover, science could and did search for correspondences between mental and cerebral states, but the analysis of the underlying substance hit against the rock of the "Unknowable," as did the analysis of the nature – though not the fact – of the correlation between its two aspects of the mental and the physical.

Spencer admitted that in science it was more reasonable to try to translate mind in terms of matter than to translate matter in terms of mind, but he denied that this was materialism. He held that any reduction of mind to brain occurred only at a "symbolic" level – the reduction of the linguistic term or symbol for mind to the corresponding linguistic term or symbol for brain. But the replacement of one symbol by another did not reach to the substance of which the symbols were mere names. Moreover, in the realm of epistemology the reverse reduction was also possible. While the materialist could argue that mind was just a product of matter in motion, the idealist could just as well argue that all that was known of matter was via the mind. Admitting that each was right as concerned part of the story, Spencer stated that philosophers, putting themselves above the fray of contending limited viewpoints, "will see that though the relation of subject and object renders necessary to us these antithetical conceptions of Spirit and Matter, the one is no less than the other to be regarded as but a sign of the Unknown Reality which underlies both." (1867: 549-50) Spencer's term for this philosophy was "symbolic realism."

Spencer was able to develop a system of evolution, lacking only an exposition of inorganic evolution. He had gone beyond Darwin's more timid and limited extension of evolution to man, mind, and morals with his distinction among the three realms of the inorganic, organic, and super-organic, though at the price of a vaguer and more abstract theory of the mechanism of evolution. Spencer was not alone in working out the implications of evolution as a scientific and philosophical world view. Darwin's main scientific supporter, T. H. Huxley, worked both to popularize evolution and to analyze its philosophical foundations.

CHAPTER THREE

T. H. Huxley: Monism and Epiphenomenalism

T. H. Huxley (1825-1895) was a prolific writer on many subjects: his *Collected Essays* dealing with the popularization, theory, and philosophy of evolution alone run to nine volumes and his scientific writings another four. His role as "Darwin's bulldog," a sobriquet earned because of his implacable defense of what Stephen Jay Gould has called the "fact of evolution," has obscured his subtle differences with Darwin as concerned the mechanism of evolution, and his independent approach to the philosophical framework for evolutionary theory.

Huxley's attitude towards evolution by natural selection was a complex one. In his 1860 review of *Origin of Species* he referred to "Darwin's hypothesis" of natural selection and compared Darwin to Copernicus, who had also formulated a new and radical hypothesis, that of the heliocentric universe.[1] Copernicus had changed the place and motion of the earth, just as Darwin had changed the status of humans on earth, but Copernicus' hypothesis itself had to be adjusted and corrected by later workers. The same might apply to Darwin:

> But the planetary orbits turned out to be not quite circular after all, and grand as was the service Copernicus rendered to science, Kepler and Newton had to come after him. What if the orbit of Darwinism should be a little too circular? What if species should offer residual phenomena, here and there, not explicable by natural selection? (1860: 78)

Huxley's attitude towards natural selection did not change substantially during later years, and though he considered there to be demonstrative evidence for the truth of evolution, he reserved final approval on natural selection as its chief factor. This critical and anti-dogmatic attitude was part of his general philosophical outlook which was informed by a skepticism that rejected absolutes. Huxley was interested in Descartes, for whom doubt as to the veracity of his own perceptions played a crucial role in setting up the *cogito ergo sum* argument, and Hume, who adopted a skeptical position concerning substance and deity. This skepticism was also expressed in Huxley's religious views, where he described his attitude as 'agnostic," a term which he coined to express his inability to either prove or disprove the existence of God. Huxley was an independent thinker, despite his role as Darwin's "bulldog," and this was most clear in his views on the philosophical framework for evolutionary theory where he initially and publicly expressed doubts about Darwin's concept of continuity.

(a) Leaps and Continuity

At the outset, Huxley shared with Charles Lyell (1797-1875) concerns about Darwin's rejection of leaps in nature.[2] In his 1860 review of Darwin's *Origin of Species*, Huxley stated:

> And Mr. Darwin's position might, we think, have been even stronger than it is if he had not embarrassed himself with the aphorism "*Natura non facit saltum*" which turns up so often in his pages. We believe, as we have said above, that Nature does make jumps now and then, and a recognition of the fact is of no small importance in disposing of many minor objections to the doctrine of transmutation. (1860: 77)

Huxley returned to this problem in the conclusion to his series of lectures to workingmen, *On the Origin of Species, or, The Causes of the Phenomena of Organic Nature* (1862), where he treated the question from the angle of the relation of structure to function. He argued that enormous differences in function might be produced by a slight modification in the structure of an organism, in the same way that a small change in the arrangement of the gears of a watch made the difference between its functioning properly and its not functioning at all. Similarly, a small difference between the brains of humans and apes made for a large difference in intelligence between these species. This brought variation at the structural level in line with the Darwinian desideratum of continuity, while apparent discontinuities were displaced to the higher functional level. Huxley considered the hypothesis of continuous evolution of species to have been confirmed by Marsh's work on the genealogy of the horse which showed a whole series of intermediary changes in structure. In "Lectures on Evolution" (1876), he considered this as "demonstrative evidence" for the fact of evolution.[3]

Though eventually Darwinian in his acceptance of evolutionary continuity, Huxley demurred at the extension of the principles of evolution to the realm of ethics, and radically limited the scope of evolution to the biological. In his 1893 address, "Evolution and Ethics," Huxley held that evolution, based on the principle of competition for survival, was inconsistent with ethics, considered as the reign of cooperation among humans. Evolution, "red in tooth and claw," could not explain the altruism and concern for others essential to ethics. Using the analogy of a garden, Huxley saw ethics as the gardener cultivating and protecting delicate but beautiful flowers, which in the absence of supervision would succumb to ruder but sturdier, and hence more biologically fit flora. Natural selection, if left on its own, would choose the strong and not the ethical. Therefore, evolution could not account for ethics, and indeed, ethics was opposed to evolution: "Let us understand, once for all, that the ethical progress of society depends, not on imitating the cosmic process, still less in running away from it, but in combating it." (1893: 82) The rejection of evolutionary ethics could not be more clearly stated.

(b) Matter, Life, and Mind

Huxley was concerned throughout his philosophical work on evolution with the problem of the relationship between matter, life, and mind. In his major writing on this subject, "On the Physical Basis of Life" (1868), he noted that science had shown that protoplasm, the 19th century term for cellular substance, was essentially composed of three component parts: carbonic acid, water, and various nitrogenous bodies. In turn, these complex molecules could themselves be broken down into their atomic components of carbon, hydrogen, oxygen, and nitrogen, so that protoplasm was the result of specific kinds of combinations of chemical elements. No vital force or entity was required. The argument was by analogy: just as there was no new substance or "aqueous principle" added to oxygen and hydrogen to produce water, so there was no new substance or "vital principle" added to carbonic acid, water, and nitrogen compounds to form protoplasm. Vital action was therefore the result of the "molecular forces of the protoplasm which displays it." (1868: 154)

Huxley took the case of water as his paradigm for the relation between a compound and its components. This example is one which will recur with great frequency in the later writings of the emergentists, though with a different analysis. Huxley noted that certain properties of water, such as its fluid state between 32º Fahrenheit and 212º Fahrenheit, differed from those of either hydrogen or oxygen at similar temperatures. But the new properties of water were nevertheless the result of the combination of the two atomic components in definite proportions, and Huxley hoped that science would eventually explain these apparently new properties of the compound as consequences of those of the parts, just as technology was able to explain the operation of an artifact given its components and their manner of combination. Such a position would be termed "reductionist" in modern terminology. The same reasoning was applied to thought as well, and Huxley concluded that it was also a product of protoplasm, that of the brain in particular: "And if so, it must be true, in the same sense and to the same extent, that the thoughts to which I am now giving utterance, and your thoughts regarding them, are the expression of molecular changes in the matter of life which is the source of our other vital phenomena." (1868: 154)

Huxley's reductionist position on the mind-brain problem was complicated by what has come to be known as his "epiphenomenalism." In 1874, Huxley produced "On the Hypothesis that Animals are Automata and its History" in which his point of departure was Descartes' hypothesis that brutes were unconscious automata. Huxley argued that Descartes was correct in treating animals as automata, since their bodily functions and activities were the result of reflexes, but wrong in denying consciousness to animals. By reserving it for humans alone, Descartes' theory required a gap between animals and humans, contradicting Darwin's principle of continuity. Accepting the principle of continuity, Huxley concluded that consciousness, since it existed in humans, must also exist, if only in more rudimentary forms, in animals as well. He noted the following arguments in support of his thesis: consciousness in humans was subject to variation by degree, the lowest of which corresponded to types of consciousness in animals; animals possessed, though in a less developed way, that part of the brain – the cerebral cortex – that was believed to be the organ of consciousness in humans; animals, though they lacked language and therefore the capacity for trains of thought, nevertheless had trains of feelings. From this he concluded that consciousness came in degrees and that animal consciousness foreshadowed that of humans.

While he believed that both humans and to a lesser extent animals had consciousness, and that this consciousness was produced by "molecular motions" in the brain, Huxley found no evidence for the causal efficacy of consciousness – its return action upon the molecular structure of the brain. He pointed to experimental evidence that decorticated frogs continued to go through many of the behavioral acts of their normal repertoire despite the impossibility of experiencing any consciousness. Huxley extended this reasoning to the normal animal, and concluded, rather rapidly and on the basis of this one example, that no mind-to-brain causality occurred. Consciousness was an accompaniment of the functioning of the brain, not an active agent. Using the analogy of a train and its steam whistle, he stated that consciousness was but a sign or epiphenomenon of the working of the brain:

> The consciousness of brutes would appear to be related to the mechanism of their body simply as a collateral product of its working, and to be completely without any power of modifying that working as the steam-whistle which accompanies the work of a locomotive engine is without influence upon its machinery. Their volitions, if they have any, are an emotional indicative of physical changes, not a cause of such changes. (1874: 240)

Using the argument from continuity once more, Huxley extended this proposition to humans as well. The molecular states of the brain, which he called "neuroses," were the cause of the states of mind, termed "psychoses." Neuroses caused psychoses, but psychoses did not cause neuroses; there was forward, but no reverse or return action. Because of this sharp distinction at the level of causality between brain and mind, Huxley has occasionally been construed as a substance dualist, while because of his denial of the efficacy of mind over body, he might also be considered as an eliminationist who rejected the value of mind. On his own view, however, he was neither. He was not a dualist, since mind and brain were aspects of the one substance he hypothetically accepted, and he was not an eliminationist, since mind, even though it could not act on the brain, did have a function of crucial importance. In humans it was the source, and indeed the only source of knowledge of reality, as well as being the site of emotional reactions to the world. Without mind, life would still be active, Huxley could argue, using the example of the decorticated frog. But, it must be added, such a life would be very boring indeed, bereft of emotions and thought. Mind as the condition for knowledge was not superfluous, and though it lacked causal powers it possessed cognitive ones of prime importance.

(c) Monism, Materialism, and Idealism

It might seem that Huxley's position was materialist, since life and mind were to be explained on the basis of matter and its modes of combination. Huxley was aware of this difficulty, and attempted to extricate himself from such a charge, a serious one in Victorian England. He argued that his philosophy would amount to a materialist analysis only if he held two further propositions: in the first place, that knowledge was absolute and not relative, and in the second place, that knowledge of cause and effect was knowledge of necessary laws and not merely knowledge of the order of succession of contingent facts. Appealing to theses of Kant and Hume for support, Huxley denied these two propositions. Invoking Kant, Huxley argued that humans could not have knowledge of the nature of matter, any more than they could have knowledge of the nature of mind or spirit. Huxley noted that matter and spirit were both "names for the imaginary substrata of groups of natural phenomena." (1868: 160) As Kant argued, humans cannot go beyond knowledge of the phenomenon to knowledge of the noumenon. Knowledge concerned the appearance of things, and could not reach the nature of the things in themselves. This was a theme also developed by Spencer in his concept of the "Unknowable."

Further, Huxley rejected the "iron law" of necessity. Necessity illegitimately went beyond the facts and the laws that link those facts. Huxley poetically put the matter as follows: "Fact I know; and Law I know; but what is this necessity, save an empty shadow of my mind's throwing?" (1868: 161) Invoking the skepticism of Hume, Huxley stated that the fundamental doctrines of materialism, like that of spiritualism or idealism, lay outside the limits of philosophical inquiry. From this Huxley arrived at the conclusion that it was indifferent whether matter was expressed in terms of mind, or *vice versa*. Both matter and mind were relative to each other, and neither was absolute in the sense of being more basic than the other.

Nevertheless, he held that a materialist terminology was preferable to a spiritualistic one, because it alone was scientifically fruitful: "For it connects thought with the other phenomena of the universe, and suggests inquiry into the nature of those physical conditions, or concomitants of thought, which are more or less accessible to us ... whereas, the alternative, or spiritualistic terminology is utterly barren and leads to nothing but obscurity and confusion of ideas." (1868: 164) However, materialistic terminology could not be extended to a materialistic philosophy, since this would be to confuse language with reality and transgress the limits of knowledge.[4]

Moreover, Huxley accepted Berkeley's proposition that the existence of knowledge apart from a thinking mind was a contradiction, and this predisposed him to epistemological idealism, where mind was the condition for the knowledge of matter. In was in this epistemological context that he made the statement, apparently the opposite of his previous endorsement of a materialist terminology: "[I]f I were obliged to choose between absolute materialism and absolute idealism, I would feel compelled to accept the latter alternative." (1871: 279) Huxley can be saved from the charge of contradicting himself only be the recognition that the domains of materialism and idealism had been sharply segregated, and that the validity of each was relative: materialism as a description of scientific objects, including the mind as an object of science, and idealism as a theory of cognition, with mind as the condition for knowledge of matter. Huxley reconciled the extremes in his adoption of monism as an ontology.

Whereas Darwin was in the privacy of his notebooks a materialist, Huxley was in the intimacy of his letters a monist. A key expression of Huxley's point of view was contained in a letter of May 1863 to Charles Kingsley. Huxley set out five ontological hypotheses about the number of "*x*'s," where *x* stood for substance: there was no *x* – atheism; there was only one *x* – materialism or pantheism; there were two *x*'s – dualism of spirit and matter; there were three *x*'s – God, soul, and matter, as in theology; and there were more than three *x*'s –

polytheism. Huxley opted for the second view – that there was just one x – as a working hypothesis and with characteristic qualification: "To say that I adopt any one of those hypotheses, as a representation of fact, would to my mind be absurd; but no. 2 is the one I can work with best." (1903, 1: 351) His position was not absolute, and he admitted that he might be grievously wrong, that there might be no substance or twenty substances. Nevertheless, the hypothesis of one substance satisfied his personal preference since it was the simplest solution "to suppose that all phenomena have the same substratum (if they have any), and that soul and body, or mental and physical phenomena, are merely diverse manifestations of that hypothetical substratum." (1903, 1: 351-351) Like Spencer, Huxley opted for a dual aspect monism, though his very recognition of a single substance underlying the correlated physical and mental manifestations was typically qualified by the hypothetical phrase "if they have any."

Huxley's view of the dialectic between materialism and idealism was based on his analysis of the history of post-Cartesian philosophy. Though he did not accept Descartes' mind-body dualism and interactionism, in "On Descartes' 'Discourse'" (1870), he recognized the French philosopher's system as the root of modern philosophy and identified two paths leading from it. The first path was based on Descartes' epistemological theory: the method of systematic doubt, the *cogito ergo sum* argument, and the principle of "clear and distinct" ideas as the criterion for truth. This was the basis for the view that mind was the condition for knowledge of reality, and led on to the idealism of Berkeley and Kant. The other path proceeding from Descartes was based on his mechanistic physiological and physical views, and led to the correlation of all the phenomena of the universe with matter in motion – materialism. Huxley traced this path to the French medical materialists and Priestley in England during the 18th century. Thus, starting from the root of Cartesian philosophy, two seemingly contradictory paths were opened up: "the one leads, by way of Berkeley and Hume to Kant and Idealism; while the other leads, by way of De la Mettrie and Priestley, to modern physiology and Materialism." (1870: 190)

It is in this context that Huxley's own paradoxical combination of descriptive materialism and epistemological idealism can be understood. The two paths were complementary, not contradictory. Legitimate materialism, as opposed to absolute materialism, was "the extension of the methods of physical science to the highest as well as the lowest phenomena of vitality." (1870: 194) This was basic if psychology and physiology were to be practiced as sciences. However, science as knowledge of reality required consciousness as its cognitive precondition, and this was what Huxley termed legitimate idealism, an idealism of an epistemological sort.

Though a popularizer of evolution, Huxley did not explicitly defend – nor reject – natural selection. Huxley agreed with Darwin on basic elements of the philosophical framework for evolutionary theory, accepting the thesis of the continuity of the evolutionary process after an initial hesitation, and arguing that properties of wholes would eventually be analyzed in terms of their parts. The situation was quite different with Alfred Russel Wallace, who defended natural selection to the point of excluding all other factors, yet differed markedly with Darwin concerning the philosophical framework for evolutionary theory.

CHAPTER FOUR

Alfred Russel Wallace: Evolution and the Supernatural

Alfred Russel Wallace (1823-1913) was the independent co-discoverer of the theory of evolution by natural selection and his paper on the subject was read to the Linnean Society jointly with that of Darwin in 1858. Nonetheless, Wallace had two significant differences with Darwin on the subject of evolution. At the scientific level, Wallace restricted Darwin's multi-factor theory of evolution to a single factor, with natural selection both necessary and sufficient to explain all evolutionary phenomena. At the philosophical level, Wallace rejected Darwin's naturalistic and quantitative framework, and advocated a supernaturalistic explanation featuring the spirit world as the origin of the qualitative novelties of life and mind. It was Wallace who was instrumental in popularizing the term "Darwinism," but the exposition he advanced was based on his own single-factor reading of Darwin's scientific theory, and his dualistic position on the philosophical framework for evolutionary theory.

(a) The Scope of Natural Selection

In *Contributions to the Theory of Natural Selection* (1870), Wallace presented a highly schematic version of the origin of species. It was one of the first attempts, if not the first, to present a deductive model of the theory. Wallace started with two propositions of biology which served as premisses to his first argument: on the one hand, unchecked reproduction would result in rapid increase in the number of organisms in all species, and on the other hand populations tended to be stationary or stable. From this he concluded that there was a struggle for existence, with deaths equalling births on the average. He then took this conclusion as the major premiss to a second deduction, adding to it the further thesis that heredity featured both a general similarity and slight differences between parents and offspring. From this he inferred that there was survival of the fittest, or what he considered as an equivalent term, natural selection: those least fit died off and those most fit survived. This conclusion then served as the major premiss for a third deduction, and was combined with the further thesis that there were changes of external conditions, due to climatic, geographic, and other environmental variations. The conclusion was that there were changes of organic forms or origin of species. Populations which survived changed in order to keep "in harmony with the changed conditions; and as the changes of conditions are permanent changes, in the sense of not reverting back to identical previous conditions, the changes of organic forms must be in the same sense permanent, and thus originate species." (1870: 166)

In this presentation Wallace made no reference to either sexual selection or inheritance of acquired characteristics as factors of evolution, even as secondary ones. He believed that sexual selection was just the differential operation of natural selection on the two sexes, and not a distinct factor of evolution as Darwin claimed. The bright colours of some male birds, rather than being a result of sexual selection and in particular, female choice, followed from the fact that females during egg-sitting were more exposed to danger than males. As a result, those females with a duller colouration and better camouflage were selected for, while the male colouration remained bright. Though Wallace was initially favourable to Darwin's views on the inheritance of acquired characteristics, and even the proposed mechanism of pangenesis, he subsequently accepted August Weismann's theory of the independence of the germplasm from the body-soma. (Weismann: 1899, 1893) In his autobiography, *My Life* (1905), Wallace listed the differences concerning sexual selection and use-inheritance as two of his four disagreements with Darwin. The other two concerned the applicability of the theory of natural selection to human evolution, and a minor disagreement on the method of propagation of arctic plants found on isolated mountain tops.

(b) The Limits of Natural Selection

The difference on the applicability of the theory of natural selection to *homo sapiens* was a major one. The background was Wallace's belief that evolution was based solely on natural selection. If any change, such as the origin of a new characteristic, occurred for which natural selection could be shown not to be responsible, then that change must be due to some extra-natural, or supernatural factor. Wallace claimed that such a situation existed for the origin of the mind and the mental abilities of humans. He believed that archaeological evidence showed the brain of early hominids to be about the same size as that of contemporary humans, an assumption which current research has shown to be false.[1] Based on that assumption, and the evidence of the lower stage of development of early hominids, Wallace posed the following problem: according to natural selection, an organ evolved when it was selected because of its use value in the struggle for existence, but much if not most of the cognitive capacity of the early hominid brain was not used. Since early hominid brains were larger than needed, and this extra capacity was useless to early man, natural selection could not have acted to develop it. Yet this same unused capacity at a later stage of development was essential for the development of civilized man's mental abilities. Wallace concluded that the musical, moral, and mathematical abilities of the mind could not have had their origin in natural selection.

As a result, Wallace stated: "We should then infer the action of a mind, foreseeing the future and preparing for it, just as surely as we do, when we see the breeder set himself to work with the determination to produce a definite improvement in some cultivated plant or domestic animal." (1870: 188) A teleological force was at work which had pre-fitted early hominids with the sorts of brain civilized humans would need only later. Whereas Darwin had argued for the analogy between natural and artificial selection in his scientific theory of evolution, Wallace argued for the analogy between natural and supernatural selection in his philosophy of evolution. Wallace did not, however, identify this superior and supernatural intelligence with deity, but instead associated it with the spirit world, of which he was a firm believer. Wallace's interest in spiritualism – table rapping, mediums, spirit-photographs, and the like – was a well developed one, as indicated in the series of articles he published devoted to this problem, "A Defence of Modern Spiritualism" (1874) in the widely read *Fortnightly Review*. [2]

(c) Qualitative Novelty and the Supernatural

Consistent with his immaterialism concerning the source of mind and the higher intellectual powers, Wallace rejected Huxley's view that protoplasm was the basis for life, and the brain the source of mind. No amount of increase of quantitative complexity in the organization of protoplasm could bring about the qualitative change involved in the origin of consciousness:

> But this greater and greater complexity, even if carried to an infinite extent, cannot, of itself, have the slightest tendency to originate consciousness in such molecules or groups of molecules. If a material element, or a combination of a thousand material elements in a molecule are alike unconscious, it is impossible for us to believe that the mere addition of one, two or a thousand other material elements to form a more complex molecule, could in any way tend to produce a self-conscious existence. The things are radically unlike, exclusive and incommensurable. To say that mind is a product or function of protoplasm, or of its molecular changes, is to use words to which we attach no clear conception. You cannot have in the whole, what does not exist in any of the parts. (1870: 365)

Wallace therefore denied the production of mind by matter, and the appearance of new properties of the brain not already possessed by inanimate matter. According to Wallace, only two possibilities remained: either panpsychism – the attribution of consciousness to all matter, including inanimate things; or dualism – the recognition of a second immaterial substance as the source of life and mind. This disjunction will be termed "Wallace's dilemma," and it will play an important role in Lloyd Morgan's thinking as he sought a way to avoid it through emergentism. Wallace formulated the choice as follows: "There is no escape from this dilemma – either all matter is conscious, or consciousness is something distinct from matter, and in the latter case, its presence in material things is proof of the existence of conscious beings outside of, and independent of what we term matter." (1870: 365) Rejecting panpsychism as contrary to common sense, Wallace opted for the dualism of matter and mind.[3]

In *Darwinism: An Exposition of the Theory of Natural Selection* (1889), Wallace argued that there were three critical moments at which supernatural forces were active. The first was "the change from the inorganic to the organic," the second was "the distinction between the animal and the vegetable kingdoms," and the third was "the differentiation of man from the brutes." (1889: 475-76) In summary, he stated: "These three distinct stages of progress from the inorganic world of matter and motion up to man point clearly to an unseen universe – to a world of spirit, to which the world of matter is altogether subordinate." (1889: 476)

In excluding the origin of life, animals, and mind from natural evolution, and in providing a causal role for the spirit world, Wallace might be considered as arguing for a saltationist and discontinuous view of evolution, in contradiction with the thesis of continuity so essential to Darwin's philosophical framework. But Wallace argued that the process of evolutionary change from animals to humans was continuous even if the source of the mental and the moral was not natural: "The point to which I wish specially to call attention is, that to prove continuity and the progressive development of the intellectual and moral faculties from animals to man, is not the same as proving that these faculties have been developed by natural selection; and this last is what Mr. Darwin has hardly attempted." (1870: 463) Wallace rejected the "fallacy as to new causes involving any breach of continuity, or any sudden or abrupt changes." (1870: 474) To the contrary, he argued that the introduction of new causes leading to life and mind could be compared to the continuous "change that takes place in the curve in which a body is moving when the application of some new force causes the curve to be slightly altered." (1870: 476) This point requires further elucidation. Is it possible to have a theory which combines evolutionary continuity with a supernatural source for novel qualities, or does the latter necessarily imply a discontinuous process and a saltationist point of view?

(d) Continuity and Discontinuity

In order to answer this question, Wallace's supernaturalist position on qualitative novelty may be contrasted with that of Joseph Le Conte (1823-1901), who argued for qualitative novelty combined with saltationist leaps in evolution. Le Conte was an American evolutionist who had been trained by Louis Agassiz, a life-long opponent of evolution. But Le Conte, unlike his teacher, accepted the theory of species transmutation and proposed a multi-factor theory, including Lamarckian, Darwinian, and what he termed "rational" factors. In *Evolution: Its Nature, Its Evidences and Its Relation to Religious Thought* (1891), Le Conte identified five "planes" of material existence. These consisted of the base plane of "elementary existence" composed of the chemical elements, followed by the plane of chemical compounds, and the planes of vegetable, animal, and human existence. In an earlier essay the "Correlation of the Vital with Chemical and Physical Forces" (1874), he argued that there were natural forces or "affinities" – physical, chemical, vital, and voluntary – whose function it was to "raise matter from each plane to the plane above, and to execute movements in the latter." (1874: 174) These forces were all natural factors, and Le Conte rejected any mysterious or supernatural sources for the new planes. Each force arose from the previous one in a step by step process:

> With each elevation, there is a peculiar force added to the already existing, and a peculiar group of phenomena is the result. As matter only rises step by step from plane to plane, and never two steps at a time, so also force, in its transformation into higher forms of force, rises only step by step. Physical force does not become vital except through chemical force, and chemical force does not become will except through vital force. (1874: 175)

Though a higher force arose naturally and step by step from the next lower, the transition was not continuous. Le Conte noted that the metaphor was not one of gradually inclined planes, but of "successive planes raised one above the other … In the ascensive scale of forces, in the evolution of the higher forces from the lower, there are places of rapid, paroxysmal change." (1874: 195) In this philosophy of evolution Le Conte combined elements of qualitative novelty – higher planes arose from lower ones, but were not identical to them – with saltationism – the higher planes were not continuous with the lower ones, but involved leaps at the points of "rapid, paroxysmal change." The acceptance of qualitative novelty in the evolutionary process and the related unpredictability of the novel was clearly expressed as follows: "[W]ith every new form of force, with every new birth of the universal energy into a higher plane, there appear new, unexpected, and previous to experience, wholly unimaginable properties and powers." (1891: 318) At the rational or fifth plane of existence, Le Conte considered that among the new properties was that of the immortality of the soul, leading to a reconciliation of evolution and religion.

Le Conte shared with Wallace the acceptance of qualitative novelty in the evolutionary process, but differed in that he did not posit a supernatural source for the appearance of new planes of existence. Le Conte, however, differed from Wallace in holding that new planes arose through saltationist, discontinuous leaps. I propose the following models to illustrate Wallace's claim that his theory did not involve a discontinuity in what may be called the graph of evolution, but only a change in direction, unlike Le Conte's saltationism, which involved gaps and leaps in the curve of evolution:

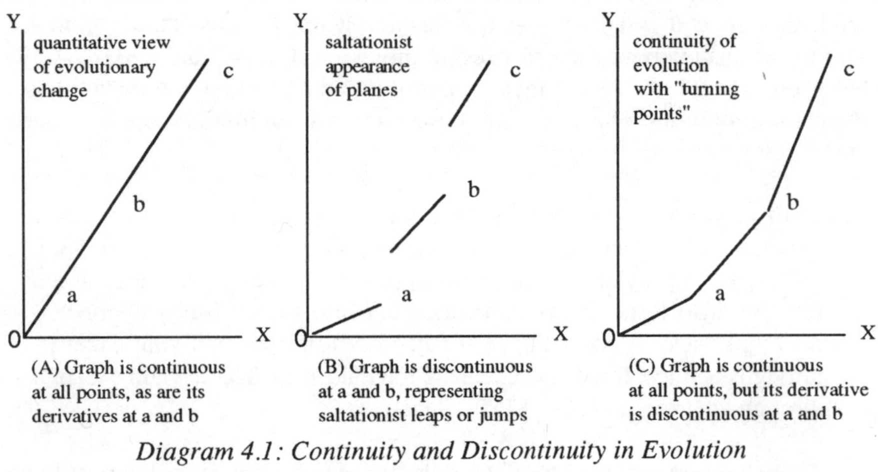

Diagram 4.1: Continuity and Discontinuity in Evolution

In these graphs, the X-axis stands for time, while the Y-axis stands for development of organic structures. Parts of the graph with constant slope represent the evolution of a single type of organic structure, so that the line *oa* stands for the evolution of life before animals, *ab* the evolution of animals, and *bc* the evolution of animals with minds. Graph *A*, which is continuous and whose first derivative is also continuous is meant to represent Darwin's point of view.[4] Change occurs by "infinitesimal" increments; there are no turning points where new qualities arise. Qualities found later in the evolutionary graph must exist earlier as well, though in lesser quantity. A saltationist theory such as that of Le Conte results in a graph as in *B* which is discontinuous at points *a* and *b*. But Wallace's theory suggests a representation as in *C* – the graph itself is continuous, but its first derivative is discontinuous at points *a* and *b*, which correspond to the points of appearance of life and mind. For Wallace, the turns at *a* and *b* were due to supernatural intervention. The emergentist point of view will see these qualitative novelties as the natural result of the increase in complexity of evolving structures.

CHAPTER FIVE

G. J. Romanes: Evolution of Mind

George John Romanes (1848-1894) was a collaborator with Darwin on instinct, and one of the co-founders, along with Darwin himself and Lloyd Morgan, of comparative psychology.[1] In his multi-volume work *Darwin and After Darwin*, the last two volumes of which were edited by Lloyd Morgan after Romanes' death, Romanes was particularly concerned to distinguish the single factor neo-Darwinism of Wallace from the richer, multi-factor theory of Darwin himself. To avoid confusing Wallace's views with those of Darwin, Romanes coined the phrase "neo-Darwinism" to designate Wallace's position. From this it might appear that Romanes was a strict Darwinian and defended Darwin's theory in an absolute sense. However, he criticized the view that natural selection was sufficient to explain species origination, and developed the theory of physiological selection to supplement Darwin's theory. Particularly in his philosophical speculations concerning monism, Romanes was the link between the 19th century debate over the framework for evolutionary theory and Lloyd Morgan.

(a) Physiological Selection

In Darwin's theory of natural selection slight morphological changes that conferred evolutionary advantage would be selected, and over a long period of time these modifications would accumulate and result in what taxonomists would classify as a new species. Romanes, however, believed that Darwin's natural selection, even when aided by sexual selection and use-inheritance, was not enough to achieve inter-population infertility of distinct species. Swamping of varieties would occur, unless the modifications were to the reproductive systems as well.

But the reproductive system was an internal physiological system, not an external morphological one upon which natural selection typically acted. Romanes' solution was to postulate a further factor of evolution, which he termed "physiological selection" so that variations in the reproductive system became conditions for the development of incipient species. The variations in the reproductive system upon which physiological selection acted were not initially useful, and they would not normally have been preserved by natural selection. But if conserved by physiological selection, and combined with other variations which "supervened" on them, the result would be the infertility of sub-populations and the divergence of their characteristics, leading to the formation of distinct species:

> When variations of a non-useful kind occur in any of the other systems or parts of organisms, they are, as a rule, immediately extinguished by intercrossing. But whenever they arise in the reproductive system in the way here suggested, they tend to be preserved as new natural varieties, or incipient species. At first the difference would be in respect of the reproductive systems; but eventually, on account of independent variation, other differences would supervene, and the variety would rank as a true species. (1897, 3: 43)

Romanes believed that Darwinian natural selection was necessary, but not sufficient to produce speciation. Unlike Wallace, who reduced all of evolution to just the chief Darwinian factor, Romanes extended Darwin's theory from natural selection to physiological selection as well.

(b) Evolution of Mind

Romanes was a founder of the comparative psychology of animals and humans, and studied the evolution of mind in both. In *Animal Intelligence* (1883), he set out his basic definition of mind, and made a detailed study of the intelligence of animals from the mollusks to the primates, exclusive of man. In *Mental Evolution in Animals* (1883) he developed his theory, particularly as it concerned the concepts of consciousness and intelligence, sensation and perception, and instinct and reason. The companion volume, *Mental Evolution in Man* (1888), extended this work to humans. Like Darwin, and unlike Wallace, Romanes argued for a naturalistic view of all aspects of human evolution. Human mental capacities were based on the foundation of emotions and intellectual capacities which were common to both humans and animals, and evolved from the latter. Romanes considered it illegitimate to limit instincts to animals and reason to man, for though the former was more developed in animals and the latter in humans, both occurred to some extent in each. Human minds arose naturally from an animal origin in the course of evolution.

Romanes included a comparative table of animal and human development, putting into correlation the phylogeny of the animal mind and the ontogeny of the human one. He divided this tree of mental life into 50 degrees, though he annotated only the first 28 with organic species from the amoeba to the great apes and domesticated dogs. Phases of human development mirrored or recapitulated the stages of animal mental development: the anthropoid apes and dogs at level 28 equalled the human infant of 15 months, while monkeys and elephants equalled the baby of 12 months. Insects and spiders were at the same psychological level as the human infant of 10 weeks, and lower forms of life corresponded to embryological stages of human development. Romanes' correspondence between the stages of mental development in individual humans and the evolution of animal mind was similar to Ernst Haeckel's theory of recapitulation, according to which ontogeny recapitulated phylogeny. (Gould, 1977b) The gradation of species in terms of intellectual powers was founded on anecdotal evidence, mainly from animal owners. It was for this reason that Romanes listed dogs above monkeys, basing himself on reports from doting pet owners with high regard for canine intelligence. One of Lloyd Morgan's tasks in comparative psychology would be to set out experimental criteria and a "canon" of interpretation to replace Romanes' anecdotal methodology.

(c) Subjective, Objective, and Ejective

In *Mental Evolution in Man* (1888), the volume dealing specifically with human mental development, Romanes posed the question as to the nature of knowledge of human and other minds. In the case of self-knowledge, he replied that humans knew their own mind subjectively or introspectively, through "an immediate cognizance of a certain flow of thoughts and feelings." (1888: 15) But knowledge of other minds was neither subjective, since not immediate self-knowledge, nor objective, since it involved more than the mere description of behaviour. Knowledge of other minds involved an interpretation of objective behaviour combined with a projection of subjective consciousness:

> Therefore, all our knowledge of mental activities other than our own really consists of an inferential interpretation of bodily activities – this interpretation being founded on our own subjective knowledge of our own mental activities. By inference we project, as it were, the known patterns of our mental chromograph on what is to us the otherwise blank screen of another mind; and our only knowledge of the processes there taking place is really due to such a projection of our own subjectivity. (1888: 16)

Romanes adopted the term "ejective" from W. K. Clifford (1845-1879), to designate "the distinctive character of a mind or mental process other than our own in its relation to our own."[2] (1888: 16) In considering the mental it was essential to distinguish those activities of bodies that were indicative of the presence of mind and those which were not. Romanes identified the dividing line as the presence or absence of choice in the behaviour of a living organism, which in turn he associated with consciousness on the subjective side. Romanes considered that real choice intervened only when an animal or human modified its behavioural repertoire in a way not explicable solely by heredity. In such cases, the organism learned from its own experience and exhibited novel activities, rather than rigidly adhering to a pattern common to all members of its species. This added an element of uncertainty since the behaviour of the animal could not be predicted simply from its previous activities. Romanes concluded: "In other words, ejectively considered, the distinctive element of mind is consciousness, the test of consciousness is the presence of choice, and the evidence of choice is the antecedent uncertainty of adjustive action between two or more alternatives." (1888: 18)

Romanes held that the physical basis of mind was the brain, and that intelligence was linked with brain size or mass, and brain structure or complexity. This was similar to Spencer's view that intelligence was associated with the quantity and complexity of the brain. For Romanes, consciousness was the product of the cerebral hemispheres, while reflex action was due to the lower ganglia. Consciousness and reflexes could be distinguished quantitatively since reflex actions required less time for execution, while conscious actions required more time. Consciousness evolved from reflex actions as organisms entered into more varied relations with the environment. Specialized reflexes became overburdened by new and complex situations, and the higher nerve centers of the cortex developed to take over the task of focusing these many and varied stimuli. This led to the "higher aptitude of discrimination" characteristic of mind.

In his analysis of the development of mental abilities, Romanes analyzed ideas into three sorts: percepts, recepts, and concepts. Percepts were ideas which were "simple, concrete, and particular," recepts were the next higher level of ideas, compounded out of percepts by the law of association into "complex, compound, or mixed ideas," while concepts were the highest ideas, being "general, abstract, or notional." (1888: 395) The intermediary category of recepts was introduced in order to provide continuity from perception through to conception, and thus continuity between animals and humans. Animals could attain the level of recepts, which were kinds of "pre-concepts," so that concepts did not arise suddenly among humans.

A similar situation obtained for self-consciousness, which as manifested through language and speech appeared limited and unique to humans. However, this would pose an embarrassing difficulty for the theory of continuity between animals and humans which Romanes, following Darwin, defended. To parry this problem, he introduced two gradations between consciousness and self-consciousness. These he termed "receptual self-consciousness" in brutes and infants, and "pre-conceptual self-consciousness" in older children: "Receptual or outward self-consciousness, then, is the practical recognition of self as an active and a feeling agent; while conceptual or inward self-consciousness is the introspective recognition of self as an object of knowledge, and therefore, as a subject." (1888: 199-200) Full conceptual self-consciousness was achieved in adult humans. Self-consciousness was not an all or nothing affair, and by introducing semi-qualitative gradations or intermediate stages, Romanes again hoped to avoid the need to recognize a completely new property of the human mind which could be explained only on saltationist grounds, as argued by St. George Mivart and others.[3] Romanes attempted to elaborate Darwin's theory of mind in such a way as to provide a smooth transition from animals to humans; hence his enthusiasm for the concept of the receptual, the existence of which was otherwise difficult to justify. The emergentist theory of mind produced by Romanes' collaborator Lloyd Morgan will eliminate the need for this transitional concept.

(d) Monism Once Again

In the essays contained in *Mind and Motion and Monism* (1895), posthumously edited by Lloyd Morgan, Romanes developed his philosophical position on the mind-body problem, in which he focused on the existence or absence of causal relationships between mind and body, rather than their status as parts of the same or distinct substances. In Romanes' view, both materialism and spiritualism admitted the reality of causal interactions between mind and body. According to spiritualism, mental changes caused material changes, while materialism held the converse view. Monism maintained that there was no causal relationship between the two, and argued that the two supposedly distinct classes of phenomena – the material and the mental – were "really one and the same." (1895: 42)

This denial of causal interaction also occurred in three other theories: pre-established harmony asserted that the association between the two types of change was pre-determined by a superior mind; chance theory held that the association of mental and material changes was random; and pure idealism denied real existence to the material realm. Romanes rejected these three theories out of hand. Pre-established harmony diverted attention from the nature of the mind-body problem to that of a superior intelligence. The chance theory was contradicted by the "intimate, constant, and detailed" association between mental and cerebral changes. Pure idealism, interpreted as a sort of solipsism, though incontrovertible by logic alone, was contradicted by common sense and its practical recognition of a material world independent of the self.

Romanes then turned his attention to the first three positions: spiritualism, materialism, and monism. He noted that the main advantage of spiritualism was epistemological. Its thesis that mind caused matter proceeded from what was immediately known – mind – to what was the object of knowledge – matter. On the other hand, if mind could act so as to cause material changes in the brain, it would have to do so by creating or destroying energy, which Romanes rejected as contrary to the principle of the conservation of energy.

The advantage of materialism lay in the fact that it was supported by physiology and physics. Romanes stated that research into the functioning of the brain had shown "the relation between neurosis and psychosis to be a constant and concomitant relation, so that the step between this, and regarding it as a causal relation, seems indeed a small one." (1895: 56) But Romanes did not make this small step, as he considered causality to be an anti-symmetric relation such that if *A* caused changes in *B*, then *B* could not have a return effect and cause changes in *A*. As a result, if brain states caused mental states, then the reverse causation would not be possible.

This led Romanes to identify materialism with Huxley's epiphenomenalism, a position which he considered inconsistent with evolution. If mind could not cause physical change, it was useless in the struggle for existence. As a result, according to the theory of natural selection, it would not be selected for and should never have evolved: "If psychosis is, as supposed, a function of neurosis, the doctrine of natural selection alone would forbid us to imagine that this function differs from all other functions in being itself functionless." (1895: 73) Romanes' concept of causality as necessarily asymmetric led him to identify materialism and epiphenomenalism in a way that a more careful analysis of causality would avoid, though his critique of epiphenomenalism on evolutionary grounds was independent of this difficulty.

Having rejected materialism and idealism, Romanes turned to monism and its position "that mental phenomena and physical phenomena, although apparently diverse, are really identical." (1895: 83) As in Spencer's philosophy, monism was conceived of as a synthesis of the reasonable aspects of both materialism and spiritualism. Monism satisfied the legitimate physical view of materialism and the legitimate psychological view of spiritualism, without the handicaps of either, in particular the problem of the interaction of distinct causal chains.

An image used by Romanes to illustrate the nature of the correlation of the physical and the mental according to monism was that of a violin being played. An observer could see the strings being plucked and hear the sound so produced. The two types of changes, visual and auditory, involved two different modes of perceiving the same thing; in the first case, through the sense of sight, and in the second case, through the sense of hearing. But neither did the sight of the thing cause the hearing of it, nor vice-versa, as they were each aspects of the same underlying event. This analogy was then applied to the case of brains and minds: "Similarly, we may suppose that a vibration of nerve-strings and a process of thought are really one and the same event, which is dual or diverse only in relation to our modes of perceiving it." (1895: 84) It was this concept of dual aspect monism which Lloyd Morgan would initially endorse in his search for a theory of the mind-body relation.

(e) Debate over the Philosophical Framework for Evolutionary Theory

Monism was the preferred ontological theory for Romanes as well as for Spencer and Huxley, though Huxley was hypothetical in his acceptance of the position, while Spencer was categorical. Darwin's materialism also posited only one substance, explicitly identified as matter, and only Wallace broke with the monistic tradition in admitting both a natural world and a distinct, non-material spirit world.

At the same time, only Wallace clearly recognized life and mind as qualitative novelties, though their source was held to be supernatural. Both Darwin and Huxley rejected qualitative novelty in nature, Darwin in arguing for a concept of continuity which admitted only quantitative change, and Huxley in holding that the progress of science would reduce properties of wholes to those of their parts. Le Conte, though he accepted qualitative novelty in his concept of "planes" of reality, did so at the price of advocating saltationism in the evolutionary process.

The positions of Spencer and Romanes were intermediary between the extremes of Darwin's restriction of change to quantity only, and Wallace's advocacy of qualitative novelty, and may be termed "semi-qualitative." Spencer introduced the important distinction of the three realms of the inorganic, the organic, and the super-organic, though he argued that the same process, that of the transformation of the homogeneous into the heterogeneous, was the basis for all three. Romanes distinguished between the perceptual and the conceptual in the evolution of mind from animals to humans, but in order to maintain overall continuity, introduced the intermediary stage of the receptual to smooth the transition.

The question could now be posed: Was it possible to combine full recognition of qualitative novelty, particularly as concerned life and mind – as Wallace did – without rejecting the continuity of the evolutionary process and its naturalistic framework – as required by Darwin? This was the basic problem that Lloyd Morgan faced in his development of emergent evolution.

PART TWO

THE ORIGIN OF EMERGENT EVOLUTION

CHAPTER SIX

Lloyd Morgan's Formative Period

Although emergent evolution was not the exclusive property of any one thinker, the comparative psychologist Conwy Lloyd Morgan (1852-1936) was the central figure in its development.[1] Lloyd Morgan's philosophical work can be divided into three periods. The first was that of his pre-emergentist writings from 1882 to 1912, during which time Darwin, Huxley, Spencer, Wallace, and Romanes were all important influences. The second period was that of the development of emergent evolution from 1912-15. The further influences of a number of theorists can be identified: J. S. Mill, G. H. Lewes, Henri Bergson, E. G. Spaulding, and Walter Marvin. Lloyd Morgan's emergent evolution was the result of a process of creative synthesis of these two groups of influences, and as a result he was the author of a new philosophy of evolution which, for the first time, featured the emergence of qualitative novelty as its central concept. In the third period from 1915 to his death in 1936, Lloyd Morgan systematically expounded and defended emergent evolution. During this time emergent evolution was a vital philosophical trend, and its basic theses were adopted and debated by a large group of philosophers and biologists.

In the first period of his professional life, Lloyd Morgan's main scientific contribution was in the area of comparative psychology, where he replaced the anecdotal methods of Romanes with experimental procedures.[2] This aspect of Lloyd Morgan's work has recently been examined by Robert Richards in *Darwin and the Emergence of Evolutionary Theories of Mind and Behavior* (1987, ch. 8).[3] In what follows the emphasis will be on the link between Lloyd Morgan's scientific and his philosophical views – his philosophy of nature – rather than his specifically scientific contributions, though relevant aspects will be alluded to where appropriate.

(a) Nature and Novelty

Lloyd Morgan developed a philosophy of nature which attempted to incorporate novelty within a philosophical framework dominated by Darwin's thesis of the continuity of evolutionary change, and Spencer's thesis of evolution proceeding from the inorganic to the organic and then the super-organic domains. The difficulties encountered in combining these two ideas were resolved only through the development of emergent evolution in the second period of Lloyd

Morgan's work. At this stage the question was posed, but the answer remained elusive.

States of Matter

In his writings on physical science, textbooks intended for a general audience, Lloyd Morgan set out, in a series of numbered paragraphs, facts and experiments about water and other physical phenomena.[4] Though these works were popularizations of science, not original contributions to it, they are of interest for the philosophical viewpoints expressed in them. Lloyd Morgan accepted the principles of continuity and gradualism, aligning himself with the traditional Darwinian point of view on the character of change in nature. In *Water and its Teachings* (1882), he expressed this position as follows:

> Thus beginning as a gas, and by a series of gradual changes, presenting nowhere any abrupt alteration of volume or sudden evolution of heat, it ends up being an undoubted liquid. ... This clearly shows that the properties of a gas can be continually and imperceptibly changed into those of a liquid (and vice-versa). ... In neither case can any breach of continuity be detected in transition. (1882: 280-284)

This may be compared with Lloyd Morgan's point of view in *An Introduction to Comparative Psychology* (1894, 2nd edition 1903), which represented an intermediary stage in the development of his thought. In accordance with Spencer's views, Lloyd Morgan held that evolution was a single, continuous process that swept through the inorganic, organic, and super-organic stages of nature. The recognition of these three domains required an admission that something new was occurring in each, though Lloyd Morgan hastened to add that this did not imply a saltationist "gap or hiatus in the ascending line of development." (1894a: 338) He analyzed the introduction of novelties under the heading of "selective synthesis," a factor or process which produced determinate products that were "special in nature," and which occurred only in the appropriate circumstances. The examples mentioned included the formation of crystals, the changes in state of water, and chemical combinations. Consideration of these phenomena introduced a difficulty for Lloyd Morgan's belief in Darwinian continuity and he was obliged to note: "There does not appear to be a gradual and insensible change from the physical properties of the elements to the physical properties of the compound, but at the critical moment of the constitution of the compound there seems to be a new departure." (1903: 342) Lloyd Morgan illustrated this problem with a graph charting the volume to temperature relationship for water in each of its three physical states, indicating breaks at the freezing and boiling points: (1894a: 339)

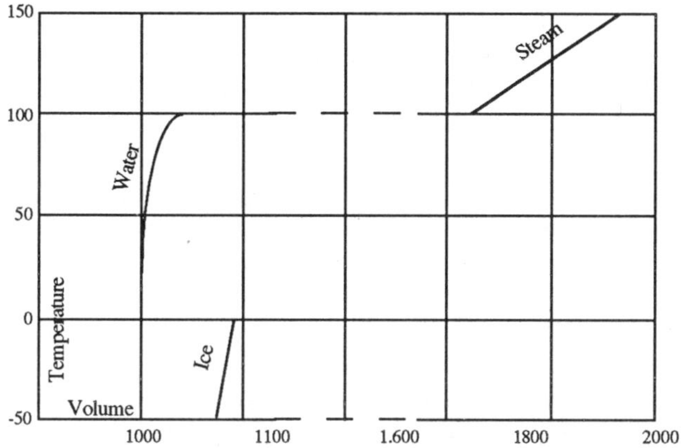

Diagram 6.1: Temperature-Volume Relations of Water

This picture appeared to contradict Lloyd Morgan's defense of continuity, since the points of "new departure" in selective synthesis – occurring at the transitions between states – seemed to involve breaches of continuity. In order to get around this difficulty, Lloyd Morgan noted that the apparent breach of continuity in the transition from liquid to vapour at 100 degrees Celsius presupposed normal atmospheric pressure. If pressure were increased, the apparent breach of continuity would no longer occur at 100 degrees, and there would be a continuous graph at that point. Of course, this just put the problem off, because at some higher temperature, boiling and the apparent breach of continuity would still occur. But his hope was that this too could be explained away as contingent and empirical:

> At the same time it should be clearly grasped that these apparent breaches of continuity are to be regarded as merely incidental to the conditions under which the phenomena are presented to our observation. The breach between the liquid and gaseous states of water holds good only under normal conditions of pressure. On these conditions it is contingent. Could we only in other matters, as has been the case with liquid and vapour through the classical researches of Andrews and others, find the appropriate conditions, every apparent breach of

> continuity would probably disappear. We are constrained to believe that evolution as a process is essentially one and continuous. By which we mean that nowhere is there evidence of supernatural interference *ab extra*. It is imperative to distinguish with due care between the results of empirical observation and their interpretation on a deeper plane of philosophical thought. The apparent breaches of continuity are empirical, and are incidental only to the limiting conditions of phenomenal presentation. (1894a: 359)

Lloyd Morgan was faced with a dilemma: Darwinian evolution required continuity, but many processes in nature seemed to involve discontinuous changes where new qualities appeared. The attempt to introduce selective synthesis as a source of novelty, with only apparent breaches of Darwinian continuity, will be discarded by Lloyd Morgan in his theory of emergent evolution. At best the concept of selective synthesis was vague, and served only to paper over the difficulties involved in combining continuity and novelty within the traditional Darwinian philosophical framework for evolutionary theory.

Organic Selection and Social Evolution

Lloyd Morgan's concern with the problem of novelty was a feature of his theory of "organic selection," which he developed in parallel to the work of James Baldwin and H. A. Osborn.[5] In "Organic Selection" (1907) Lloyd Morgan summed up his theory. He agreed with Weismann that the inheritance of acquired characteristics was not a direct factor of evolution, but argued that acquired modifications could serve as an indirect factor. He represented by *V* random variations, and by *M* modifications acquired during the life-time of an individual. In this symbolism, +*V* represented variations that were above the statistical mean for the population, and -*V* ones that were below, and similarly for +*M* and -*M*. As a result, there were four possible combinations: +*V* with +*M*, +*V* with -*M*, -*V* with +*M*, and -*V* with -*M*.

Assuming that the variations and modifications were helpful to the survival of the individual, those in group +*V* with +*M* would tend to survive, those in group -*V* with -*M* would not. Mixed results would occur for those in the other two groups. Lloyd Morgan's point was that even if +*V* was not yet the independent object of natural selection, individuals possessing it along with +*M* would be better adapted to survive. As a result the +*V* factor would tend to be inherited by offspring, even if, as Lloyd Morgan assumed, +*M* were not inherited as well. This would help the $+V$ factor during at time when it was not sufficiently developed on its own to be directly favoured by natural selection:

> The favourable variations will be inherited; and this because they are supported by the coincident favourable modifications. On this view, acquired modification is the nurse, not the parent, of variation. (1907: 69)

Organic selection could be useful in the development of complex structures, the intermediate stages of which were not sufficiently developed to serve the function eventually associated with the developed structure. This contrasted with Darwin's theory of the origin of complex structures, where the intermediate stages were preserved by natural selection because they were useful for a series of temporary functions, each of which was itself helpful for survival. Using a contemporary example, primitive structures which later evolved into wings might be initially selected for thermoregulation, then parachuting, followed by gliding, and only later because of their function of flight.[6] Use-inheritance was not required in this process.

However, in Lloyd Morgan's theory of organic selection there was an inner relation, or organic relation, between the two factors of natural selection and acquired modifications. They acted jointly and the net result was the evolution of complex forms, the intermediate stages of which were not viable even for temporary functions. The concept of organic selection bore a resemblance to Romanes' physiological selection, in that both postulated mechanisms for the preservation of characters which were not sufficiently developed to be preserved by natural selection, but which when further developed would be preserved and would contribute to the origin of new species. The concept of organic selection, though championed by Lloyd Morgan throughout the decade of the 1900s, had little subsequent echo in the literature. Nonetheless, it indicated an attempt to deal with novelty in the scientific area of biological evolution.

Lloyd Morgan also attempted to deal with the implications of evolution in the social domain, and once again the problem of continuity and novelty were present. He distinguished between organic evolution, where natural selection operated to coordinate the organism to a physical and biological environment, and mental evolution in *homo sapiens* as a social being, where the question was that of the coordination of mind to an environment of ideas and ideals: "In social evolution among human beings, natural selection ceases to be the dominant method of evolutionary progress; and mental evolution, involving other principles, not less natural than those which the study of organic development has disclosed, becomes the central feature of the process." (1894a: 81-82)

Lloyd Morgan argued that imitation and intelligence replaced natural selection as the main factor of social evolution, and that the progress inherent in this process was consistent with continuity: "Imitation supplies the element of continuity; intelligence, that of progress." (1900c: 431) Average intelligence might even fall as civilization protected the less intelligent through welfare measures, but so long as the social environment was under the lead of the most intelligent, social progress would continue. The motor of progress was not the average human being, but the achievements of the best of humanity. Social evolution thus proceeded on a different basis, or with different factors, than organic evolution; it was super-organic in the sense of Spencer. It was only in organic evolution that the fit survived at the expense of the less fit. In social evolution, the fittest played a role in "raising the level of the less fit." (1900c: 438) Again, Lloyd Morgan focused on the novel in the context of evolutionary continuity. His theories of selective synthesis, organic selection, and social evolution led up to his major work in comparative psychology, for which he is best known. (Richards, 1987)

Intelligence and Reason

In his work in comparative psychology, Lloyd Morgan examined the relations between instinct and intelligence, sense and reason, and consciousness and self-consciousness. In *An Introduction to Comparative Psychology* (1894) he distinguished between instinctive and intelligent behaviour by observing that instinctive behaviour involved no variability in action, while variable responses required the control and guidance of intelligence. Much of Lloyd Morgan's own research involved chicks and their pecking behaviour. His conclusion was that while pecking behaviour was innate, present in chicks immediately after birth, the honing of that behaviour to achieve success in feeding required practice and a degree of intelligence. Intelligence in animals, though based on experience, required that the animal be able to associate impressions gained through experience and ideas formed in its mind.

Closely linked to the concept of intelligence was that of consciousness. Though Lloyd Morgan stated that the term could not be defined, he nonetheless believed that the phenomenon could be analyzed based on the metaphor of a wave, which he represented as follows: (1894a: 18)

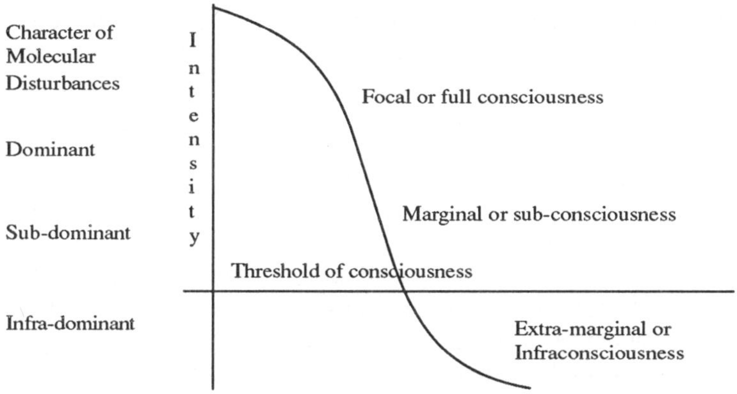

Diagram 6.2: Lloyd Morgan on the Wave of Consciousness

In a wave of consciousness the field of psychological activity was divided into the focus and the margin of attention, and this latter was further subdivided into the sub-conscious and the infra-conscious. Sub-conscious processes were just outside the focus of attention and could be brought into consciousness by a change of focus. But below the threshold of consciousness, mental states were infra-conscious and could not be brought into focus by any amount of mental effort. Dominant molecular states of the brain corresponded to focal consciousness, sub-dominant ones to marginal or sub-conscious states of mind, while infra-dominant states of the brain were correlated with extra-marginal or infra-conscious states of mind.

Consciousness played no active role in instinctive behaviour, though it might be present, and here Huxley's concept of mind as an epiphenomenon truly applied. In this case consciousness was, in Lloyd Morgan's words, "a byproduct, with no bearing whatever on the performance of the activity." (1894a: 206) But at a more developed stage of behaviour consciousness played a role to select adequate responses and direct muscular activities to specific ends. Consciousness occurred in animals with developed brains – the mammals and the birds – though Lloyd Morgan also recognized it, but to a lesser degree, in reptiles, amphibians, and fish. Though consciousness was therefore common to both humans and animals, Lloyd Morgan made a distinction between perceptual consciousness in animals and conceptual consciousness in humans. While an animal could choose and was probably conscious of choosing, humans could also reflect on their choice and imagine alternatives under the same or different circumstances. The conceptual consciousness of the human mind included reference to the self. No animal possessed self-consciousness, even though it might be conscious of its own pain or suffering. Self-consciousness was defined as involving "first, the conception of the subjective as distinguished from the objective; secondly, the concentration of the net result of all subjective experience into one generalized concept; and thirdly, the further conception of this net result as due to the determinate working of an activity which is synthetic and selective." (1894a: 321)

While Lloyd Morgan admitted that both animals and humans used inference, he also distinguished between conceptual inference or reason, and perceptual inference or intelligence. Reason stood alone as something achieved by humans and not by animals. It involved the understanding not merely of things but also of their relations, and was characterized by the methodic or systematic search for a solution, not mere trial and error. Children were still at the trial and error stage; animals never left it. Only adult humans used reason and were capable of what Lloyd Morgan called an "ideal construction" – the transformation of experience into an abstract schema. This was reflected in what has been termed "Morgan's Canon," the methodological injunction to interpret animal behavior in lower or perceptual, rather than in higher or conceptual terms:

> Once more, therefore, I must draw attention to the canon of interpretation adopted at the outset of our inquiries concerning other minds than ours, namely, that in no case is an animal activity to be interpreted in terms of higher psychological processes, if it can be fairly interpreted in terms of processes which stand lower in the scale of psychological evolution and development. (1894a: 292)

Lloyd Morgan's concept of an evolutionary transition from the perceptual to the conceptual was less smooth than Romanes', where there was the intermediate category of the receptual. Mind in its developed, adult human form involved both consciousness and self-consciousness, reason as well as intelligence, marking the novelty of the human mind relative to the animal one. But mind was an abstract concept, and Lloyd Morgan's analysis of it went beyond comparative psychology to philosophy. Like his predecessors Spencer and Romanes, Lloyd Morgan adopted neutral monism, and he was immediately faced with the problem of reconciling the apparent and pervasive novelties which he had identified in nature, especially that of the human mind, with a philosophy which was satisfied with the mere correlation of the physical and the psychical as dual manifestations of a single underlying substance.

(b) Neutral Monism

Lloyd Morgan's first systematic work in the philosophy of nature was *Springs of Conduct: An Essay in Evolution* (1885), which set out his defense of monism as an ontology. His motivation in adopting this philosophy was a familiar one in 19th Century Victorian thought – that of avoiding what were perceived as the twin excesses of both materialism and idealism, while retaining the advantages of each. Lloyd Morgan recognized that materialism had a practical advantage in scientific investigation since it did not require a spiritual substance whose interaction with matter, or creation of it, was difficult if not impossible to explain. Nonetheless, idealism had a speculative advantage, since it recognized the fact of consciousness as basic, in accord with "intuitive feelings." A materialist standpoint was useful in physiology, an idealist one in psychology, a thesis already formulated by Huxley, Romanes, and Spencer. Monism combined the best of both worlds, and according to Lloyd Morgan, was the only philosophy capable of "rendering conceivable the concomitant evolution of mind and body." (1885: 208) Lloyd Morgan expressed his dual aspect philosophy of mind using Huxley's terms of "neurosis" for brain processes and "psychosis" for mental states:

> The parallelism between neurosis and psychosis is merged in identity. They are not parallel series which run side by side, but one series which we regard under different aspects. To use the old philosophical phraseology, there are not two substances, a substance of matter and a substance of mind, but one substance, the substance of being. (1885: 208-209)

Mental and Physical

Mental and physical were as concave to convex, two sides of the coin of the same neutral substance, a metaphor due to G. H. Lewes, who also defined the term "emergent." Lloyd Morgan was familiar with Lewes' work and referred to it frequently, though his adoption of the concept of emergence would require some thirty years of further reflection.[7] Nonetheless, Lloyd Morgan did use the term "emerge" in the course of the distinction he made between psychoses and "hypo-psychoses." These latter were defined as "submerged feelings which correspond on the subjective side to neuroses, but which do not see the light and emerge in consciousness." (1885: 190) They corresponded to what he elsewhere termed the "sub-consciousness" or "infra-consciousness". Mind was the sum total of psychoses and hypo-psychoses, the system of submerged feelings and feelings which have emerged into consciousness.[8]

In *Animal Life and Intelligence* (1890), Lloyd Morgan attempted to analyze the novelty of mind and its associated qualities of consciousness and intelligence within his neutral monist ontology, while criticizing Wallace's dualism and supernaturalism. Wallace had argued that a solution to the mind-body problem required a choice between panpsychism and supernaturalism – what has been earlier termed "Wallace's dilemma." Lloyd Morgan rejected Wallace's supernaturalism, and argued that evolution advanced from unconscious life to conscious life without either the admission of a distinct and separate psychic substance, or the recognition of supernatural intervention in the otherwise natural course of evolution.[9]

Though such a position excluded Wallace's spirit world, without further qualification it led back to the other horn of Wallace's dilemma – panpsychism. According to the ontology which Lloyd Morgan accepted, it was not possible to deny some psychic aspect to lower organisms and even inanimate objects, since all things, as manifestations of the underlying substance, had both mental and physical aspects. But Lloyd Morgan recognized the difficulty in endowing lower organisms or inanimate matter with the same sort of consciousness as humans. Because such a position was untenable, contradicting both common sense and scientific knowledge, he attempted to avoid a vulgar form of panpsychism by introducing concepts to distinguish successive stages in the development of consciousness.

Kinesis and Metakinesis

Lloyd Morgan analyzed the physical side of biological evolution into three distinct stages: firstly, physiological modes of organization preceding brains, secondly, simple types of brain organization, and thirdly, complex types of brain organization. This last step posed no problem for dual aspect monism, since human mind with developed consciousness and self-consciousness corresponded to it, nor did the second step, as this was correlated with the perceptual consciousness of animals. The first step, however, was problematic. What sort of mental characteristics or states of consciousness corresponded to simple nervous systems below the level of brains? In order to reply to this question, Lloyd Morgan introduced the concepts of kinesis and meta-kinesis. All physical processes, including the physiological, were explicable in terms of energy, and for this the term "kinesis" was used. The term "meta-kinesis" was then defined as the concomitant mental manifestation of kinesis.

Metakinesis included both mental states corresponding to developed brain states, and "psychic" but pre-mental states corresponding to physical processes below that of the brain. There were metakinetic manifestations or aspects of all things, including inanimate objects, but the kinesis of the inanimate object being less than that of the animal or human brain, its corresponding metakinesis did not attain the form of mind. Lloyd Morgan did not specify what form it did take and simply noted that humans could not reproduce primitive states of metakinesis in their own developed metakinetic minds. The whole system was summed up as follows:

> I make, therefore, the following assumptions: First that there is a noumenal system of "things in themselves" of which all phenomena, whether kinetic or metakinetic, are manifestations. Secondly, that whenever in the curve of noumenal sequences kinetic manifestations (convexities) appear, there appear also concomitant metakinetic manifestations (concavities). Thirdly, that when kinetic manifestations assume the integrated and coordinated complexity of the nerve processes in certain ganglia of the human brain, the metakinetic manifestations assume the integrated and coordinated complexity of human consciousness. Fourthly, that what is called "mental evolution" is the metakinetic aspect of what is called brain or inter-neural evolution. (1885: 470)

Combined with his earlier distinctions between consciousness and infraconsciousness, this suggested to Lloyd Morgan the following representation of the process in ontogenetic or developmental terms: (1885: 173)

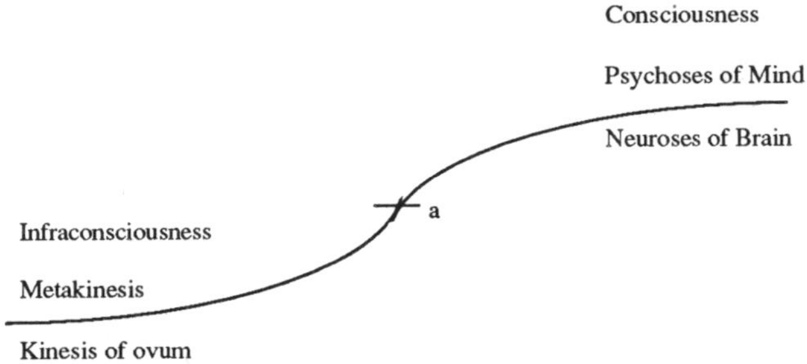

Diagram 6.3: Lloyd Morgan on Ontogenetic Development

The lower left hand of the diagram corresponded to the earliest stage of ontogenetic development, when the fertilized ovum of an animal had not yet formed a nervous system. The mental states concomitant with that primitive physiological condition were referred to as metakinetic or infraconscious. The upper right hand of the diagram represented a fully developed individual, with mental states or psychoses corresponding to brain states or neuroses. Here the individual had achieved consciousness. Point *a* was the midpoint of the curve of development, representing the moment when infraconsciousness first became consciousness. Lloyd Morgan noted that it was frequent in science to state that at point *a* the physical development of the brain had generated, or called forth, states of consciousness or mind. But, he continued, this was philosophically misleading, violating the neutral monist ontology which segregated the physical and the mental so that, accepting Romanes' concept of causality, the former could not "cause" the latter: "No conceivable increase in the orderly complexity of the molecular vibrations of brain tissue could give rise to that consciousness which differs *toto caelo* from any manifestation of energy." (1885: 173)

Nevertheless, Lloyd Morgan admitted that scientists who arrived at such a conclusion were "practically sound because they are still dealing with the same developmental curve." (1885: 194) This led him to make the following comment, of interest for its use of the term "emerge": "When they say that consciousness emerges from the physical conditions at "a', they presumably mean that at this point we are first justified in speaking of consciousness in the subjective aspect in anything like a human sense." (1885: 194) But he went on to note that this was just a convenient way of speaking, not a real ontological process.

What Lloyd Morgan had developed to this point was a rather complicated, indeed, overly complicated set of terms, none of which would be retained in his mature philosophy. Yet the concern which prompted him to coin these terms was basic to his later philosophy: the existence of qualitative novelty in the evolutionary process. It was a measure of the difficulty involved in including this proposition in a Darwinian framework that Lloyd Morgan was prompted to develop these terms in an attempt to solve a theoretical difficulty which ultimately could only be resolved by going over to a new philosophical framework.

The concept of pre-cerebral kinesis in organisms posed no problem, since it referred to a physiological process to which no conscious mental state corresponded in an organism which had not evolved a brain. But the corresponding concept of the pre-mental metakinesis of organisms, though it was logically coherent, was intuitively opaque and incapable of scientific specification. To what did pre-mental metakinetic states refer? The problem was basic to the neutral monist framework, for so long as there must be a universal correspondence between the mental and the physical, there must also be this non-empty category of the metakinetic below the level of mind. Though Lloyd Morgan identified this with infra-conscious and sub-conscious mental states, these were also mere concepts without definite content. In fact, Lloyd Morgan admitted that there was no way that conscious minds could clearly grasp the idea of what such a pre-mental psychological state might be.

A Cartesian would reject this concept, since no "clear and distinct" idea of it could be formulated. Without sanctioning Descartes' theory of innate ideas, there is merit to the rejection of concepts which are inherently incapable of exact specification, especially in a philosophy which claimed to be based on science. The further extension of the concept of metakinesis to non-living matter was a form of panpsychism. It was saved from the most egregious excesses of that doctrine by a recognition that in this case the psychological was not self-conscious or even conscious. But then what was it? Once more, Lloyd Morgan used a term – the "metakinetic infraconscious" – for which a reference could not be clearly specified.

Lloyd Morgan had placed himself before a dilemma in that either the mental appeared only with certain forms of organized matter – the brain – and was a novelty in the evolutionary process, or there was no novelty of this qualitative sort, simply quantitative degrees of metakinetic development at different moments in the evolutionary process. The first limb of the dilemma recognized qualitative novelties, the second did not. In the first case, evolutionary philosophy must go beyond Darwin's quantitative framework. This Lloyd Morgan was not yet ready to do. In the second case, qualitative change in the evolutionary process must be rejected. Lloyd Morgan, however, had come to recognize the importance of evolutionary novelties. He was impaled on the horns of a dilemma of his own making.

Physical and Metaphysical Causation

But such was Lloyd Morgan's commitment to dual aspect monism as an ontology that he persisted in its defence despite these difficulties, and applied its basic idea of coupled and complementary aspects in other areas of philosophy as well, including that of physical and metaphysical causation. In science the mind proceeded from the observation of physical processes to the formulation of theoretical conclusions. But constructs in science, including law statements such as the law of gravitation, were based on a limited number of observations. How then was the universal form of law statements justified? In "Causation: Physical and Metaphysical" (1898) Lloyd Morgan responded: "The answer is that we carry our law to an ideal limit unattainable by sense and by practical measurement. We trust to a reality of thought which we believe truer and wider than the realities of sense."

(1898a: 237) This was a first ideal construction, that of the physical sciences, to be followed by a second, metaphysical one.

A law statement, as an ideal construction, encompassed the events that fell under it in a universally quantified cause-effect statement, but Lloyd Morgan held that this did not provide an explanation in the sense of an "ultimate, underlying cause." There must be something beyond the law statement which provided this explanation, and this involved reference to metaphysical causation as dual to physical causation: "Let us speak of physical or scientific causation which refers events to their antecedents, generalizing the results of observation in an ideal scheme of physical science; and let us speak of metaphysical causation which seeks to get behind or beneath phenomena and to give the *raison d'être* of their being, generalizing its conclusions in an ideal scheme of metaphysical interpretation." (1898a: 240) As a result of the construction of these ideal schemes there was a continuity from experience to science, and from science to metaphysics.

Metaphysical assumptions were requirements of human thought, but they had to be acknowledged as assumptions, not mere generalizations from experience. Foremost among these assumptions was the recognition of the existence of a material universe independent of human experience, and the recognition of the existence of a first cause or supreme being different from the natural world. The former was required for science to be possible, and the second was the basis for religion, considered as a supernatural domain dual to the natural.

Naturalism and Supernaturalism

In "Naturalism" (1896), Lloyd Morgan argued that the natural and the supernatural were yet another couple of correlated or dual aspects of a common and underlying unity: "On this view naturalism and supernaturalism are not antagonistic, but reciprocal; do not dwell apart but are coextensive; do not deal with separate spheres, the one knowable and the other unknowable, but constitute the diverse aspects of that body of belief which is the outcome of human experience at its best." (1896c: 84) Lloyd Morgan's claim for the supernatural was only in apparent contradiction with his critique of Wallace's supernaturalistic explanation for the origin of life and mind. Lloyd Morgan's concept of the supernatural was that of deity as first and final cause, not the supernatural as intermittent and intervening cause, as Wallace argued in his theory of the spirit world as the source of life and mind. Lloyd Morgan rejected any supernatural origin for particular natural phenomena, even if deity was the ultimate metaphysical source of the natural world as a whole.

In contrast to Wallace, Lloyd Morgan argued that the scope of naturalism was universal, applying to all of the inorganic, organic, and human aspects of reality: "We must extend our conception of naturalism so as to include a naturalistic interpretation of nature in all its wealth of moods, neither excluding inorganic nature on the one hand nor human nature on the other hand." (1898a: 249) Naturalism viewed the world as cognizable under definite laws, which extended as far as the highest ideals of ethics and aesthetics. Science dealt with the natural aspect of things, religion with the supernatural in a worldview aimed at reconciling the two. Lloyd Morgan summed up his concept of naturalism, with this added component of the concomitance of the natural and the supernatural, as follows:

> 1. Naturalism sweeps through the whole range of the knowable. 2. It is nowise antagonistic to or exclusive of supernaturalism. 3. It takes as its criterion of reality direct experience prior to the analysis of science. 4. Of the two aspects of experience which primary analysis first discloses, the objective and the subjective, it regards both as strictly co-ordinate reality. 5. It asserts the inherent and intrinsic worth and dignity of the human ideals. 6. It refuses to admit that natural selection, potent as this may be as a factor in organic evolution, is to be regarded as chief naturalistic factor in human evolution. (1896c: 90)

In arguing for neutral monism, Lloyd Morgan postulated a noumenal substance which underpinned the dual aspects of the mental and the physical. In his discussion of the supernatural he introduced a second unifying entity – deity – which overarched the dual aspects of the mental and the physical. This double encompassing of the mental and the physical by noumenal substance and supernatural deity played an essential role in Lloyd Morgan's philosophy of evolution.

(c) Philosophy of Evolution

The combination of Lloyd Morgan's dual aspect monism and his views on evolution was summed up in "The Philosophy of Evolution" (1898), the key article in a series of papers published in the journal *Monist*.[10] Basing himself on Spencer, Lloyd Morgan stated that "The root-ideas of the conception of evolution are, first differentiation, and secondly the interaction of the differentiated products." (1898b: 487) Initially, noumenal reality differentiated into the dual aspects of the non-self and the self. In a second stage, these two end-products of differentiation interacted, and through integration produced a "new and more complex unit" –

experience. In turn experience differentiated into object and subject. As a result, the distinction between self and non-self was more fundamental than that between subject and object, since the former depended only on the differentiation of noumenal reality, whereas the latter depended on the further integration of the self and the non-self into experience. Lloyd Morgan illustrated this abstract schema with the following diagram:

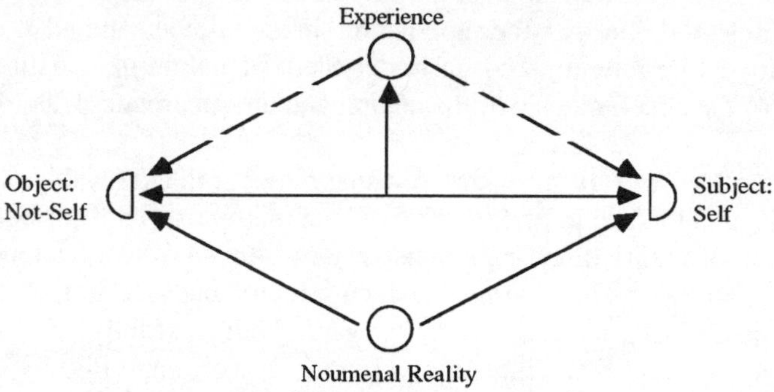

Diagram 6.4: Lloyd Morgan on Analytic Monism

The diagram was accompanied by the explanation:

> The lower circle represents the monistic unity prior to dualistic differentiation; hence arises the self on the one hand and the not-self on the other, which by their interaction give origin to experience, including both sense and thought. The dotted lines express the analysis of experience into object and subject coterminous with the not-self and the self of differentiation. But whereas the subject and object of analytic dualism are dependent on the occasionalism of experience, the self and not-self are persistent so long as the differentiation holds. (1898b: 500-501)

When this model was applied to the cosmos as a whole, the noumenal reality was referred to as the basic "activity" or "source" of its evolution. This was the cause, in the metaphysical sense, of the differentiation and interaction of the cosmic elements, which in the conditions of time and space, brought about the world as an existent. This metaphysical cause was described as timeless, spaceless, eternal, infinite and transcendent of experience. Its nature could not be fully described: "[I]t must suffice to say that the underlying metaphysical activity, as cause, is neither the product of evolution nor its precursor in time; it is that timeless omnipresent existence in and through which evolution is rendered possible." (1898b: 501) This corresponded, in Lloyd Morgan's reconciliation of science and religion, to God, and would later be associated with Samuel Alexander's concept of "*nisus* towards deity."

In discussing evolution in organic and inorganic nature, Lloyd Morgan made the distinction between intrinsic and extrinsic differentiations. In an extrinsic differentiation, a combination of external factors acted to produce an effect, such as planets circling the sun under the joint influence of gravitational attraction and inertial velocity. Here the result was predictable given knowledge of the governing factors and initial conditions. But with intrinsic differentiation, as instanced in the regular pattern displayed by a growing crystal, or the compounds formed by chemical combination, Lloyd Morgan stated that the result was not predictable: "No one, in the absence of observation or analogy based on practical experience (his own or that of others), could foretell what new characters a compound, resulting from the chemical union of well-known elements, would possess." (1898b: 489) This concept of the unpredictability of the novel will re-appear in Lloyd Morgan's later emergentism.

In this formative period of his philosophical work, Lloyd Morgan had developed two sets of concepts. The first set included the correlated aspects of material and mental, objective and subjective, physical and metaphysical, scientific and religious, and natural and supernatural – all of which were essential components of his analytic or dual aspect monism. A second set of concepts comprised those of selective synthesis and apparent breaches of continuity, psychosis and hypo-psychosis, kinesis and meta-kinesis, consciousness and infra-consciousness. These were developed in an effort to situate novelty, that of the mental in particular, within a monistic framework, but were at best *ad hoc* concepts that never integrally fit into the overall structure of his philosophy.

This difficulty was symptomatic of the problems involved in combining a neutral monist ontology, or theory of being, with an evolutionary ontogeny, or theory of becoming. The ontology postulated a timeless substance which a-causally bifurcated into the two correlated aspects of the physical and the mental. But the ontogeny posited a temporal process of change whereby the physical evolved until it reached a state, that of the brain in animals, to which conscious mind corresponded. The neutral monist ontology and the evolutionary ontogeny were in conflict, though Lloyd Morgan in his philosophy of evolution attempted to reconcile them. The development of emergent evolution would eliminate the need for the *ad hoc* concepts, once qualitative novelty was recognized as integral to the evolutionary process, though the status of neutral monism and its correlated aspects would still pose a problem.

CHAPTER SEVEN

Influences on Lloyd Morgan

The development of emergent evolution involved the combination of two ideas: the first held that evolution was a general phenomena, sweeping through all domains of nature, while the second stated that at specific points of development, new levels of organization appeared, featuring novel qualities. Most theoreticians of the Darwinian tradition had opted for a mode of evolution characterized by quantitative or at best semi-qualitative change. Qualitative novelty had been fully accepted only by Wallace, but he had postulated a supernatural source for it. Moreover, qualitative change in species production was associated with the saltationist and anti-Darwinian views of Argyll, Owen, and Mivart. Le Conte, who came closest to an emergentist theory with his distinction between planes of reality, nonetheless adopted a saltationist view of the transition from one plane to another. As a result, the recognition of qualitative novelty in the evolutionary process, essential to the development of emergent evolution, was regarded with suspicion: either it was seen as a concession to saltationism, or as a concession to supernaturalism, both of which were unacceptable from a strict Darwinian point of view.

Consequently, it was not a simple matter for Lloyd Morgan to move to emergent evolution. The following section will trace how he did so. An examination of Lloyd Morgan's published texts, along with his notes of the period 1912-15 in the Lloyd Morgan Archives at the University of Bristol, identifies the following major influences on Lloyd Morgan's thought: the English philosophers J. S. Mill and G. H. Lewes, the French philosopher Henri Bergson, and the American "New Realists" E. G. Spaulding and R. W. Marvin. The work of these philosophers forms a second background – supplementary to that of Darwin, Spencer, Wallace, Huxley, and Romanes – to the development of emergent evolution by Lloyd Morgan.

(a) Resultant and Heteropathic Laws: J. S. Mill

J. S. Mill's (1806-1873) *A System of Logic* first appeared in 1843 and went through numerous editions up to the last one of 1872.[1] In the section dealing with causality, Mill formulated the distinction between heteropathic laws and laws of composition. Like Hume, Mill rejected any concept of cause as a power to bring about or produce its effects. Rather, he was concerned solely with the orderly succession of phenomena. Mill then noted: "To complete the general notion of causation on which the rules of experimental inquiry into the laws of nature must be founded, one distinction still remains to be pointed out: a distinction so radical, and of so much importance, as to require a chapter to itself." (1872: 370)

The distinction was that between two types of causes, the mechanical and the chemical. Mill first considered the ordinary case of mechanical causation where the combined effect of two or more causes acting jointly could be calculated given knowledge of their individual effects when acting singly. The example given was the familiar one of two physical forces acting upon a body. Given previous knowledge of the magnitude and direction of the individual forces, the magnitude and direction of the resulting force could also be calculated, by representing each force as a vector. The result of the joint action of the two forces was then given by the diagonal of the parallelogram formed by the two vectors. Mill termed causation of this type "composition of causes." Though this type of causation applied to mechanical problems, Mill believed that this was not the case for chemical and biological phenomena. He was bold in his claim for water:

> The chemical combination of two substances produces, as is well known, a third substance with properties different from those of either of the two substances separately, or of both of them taken together. Not a trace of the properties of hydrogen or of oxygen is observable in those of their compound, water. (1872: 371)

The same applied to biological entities:

> All organized bodies are composed of parts similar to those composing inorganic nature, and which have even themselves existed in an inorganic state; but the phenomena of life, which result from the juxtaposition of those parts in a certain manner, bear no analogy to any of the effects which would be produced by the action of the component substances considered as mere physical agents. (1872: 371)

Mill stated that actions of this sort were due to laws which he termed "heteropathic."[2] Mechanics, because its laws were subject to the composition of causes, could be formulated as a completely deductive science; not so chemistry or biology, since their laws could not be deduced from knowledge of the laws of action of their component parts. However, once the non-deducible or heteropathic laws were accepted as basic postulates or axioms of the sciences of chemistry and biology, the rest of the phenomena described by these sciences could then be deduced.

Though Mill affirmed the heteropathic status of the laws of chemistry and biology, he was ambivalent as to the status of mental laws. He noted that physiological causes sufficed for the explanation of sensation, and believed that this was probably, though not certainly, the case for cognition as well. He held that it was theoretically possible that psychology could be reduced to physiology, were corresponding physiological states to be found for thoughts as they had been for sensations. Psychology was a distinct science only provisionally, because this reduction had not yet been accomplished. Heteropathic laws were certain only for chemistry relative to physics, and biology relative to chemistry.

Mill's discussion of heteropathic laws was limited in extent to a few pages in a thousand page volume, and was peripheral to the main concerns of his work. Moreover, though he admitted the probability of evolution being true – in a footnote, not in the main text (1872: 328, footnote 6) – he was not concerned with the relevance of evolution to heteropathic laws.

(b) Emergents and Resultants: G. H. Lewes

G. H. Lewes (1817-1878) presented his ideas on novelty in his multi-volume *Problems of Life and Mind* (1874-1879).[3] Unlike Mill, whose heteropathic laws were based largely on considerations of chemistry and biology, Lewes developed a more general theory of what he termed "emergents." Indeed, it is surprising to see to what extent Lewes foreshadowed Lloyd Morgan. Lewes accomplished much more than merely renaming the effects of Mill's heteropathic laws as emergents, though he did not fully succeed in combining his philosophy of emergence with the concept of evolution.

In *Problems of Life and Mind* (second series, 1877) Lewes situated biology in its relation to the other sciences. Following Comte's classification of the sciences, he placed biology after mathematics, astronomy, physics, and chemistry. Each science added a new method to the preceding. Mathematics began with abstraction, deduction, and definition. Astronomy added observation, physics added experiment, chemistry added nomenclature, and biology contributed classification. This hierarchy of sciences was completed by sociology and psychology.

Consistent with the monism he shared with Spencer and others, Lewes rejected the notion of a vital force or principle in biology, just as no special force or principle was required to explain the growth of crystals. Yet he recognized that life could not be confounded with the chemical compounds which produced it. Biology differed from physics and chemistry in that biology considered organisms as a whole in their relation to their environment or milieu, while physics and chemistry studied parts of the organism in abstraction from the whole and its surroundings: "The biologist will employ chemical and physical analysis as an essential part of his method; but he will always rectify what is artificial in this procedure, by subordinating the laws of Physics and Chemistry to the laws of Biology revealed in the synthetic observation of the organism as a whole." (1877: 14)

In *Problems of Life and Mind* (third series, 1879), Lewes further addressed the problem of qualitative novelty. He rejected two claims often made by non-emergence theories: firstly, that what occurred later in the evolutionary process must reproduce the general properties of what had occurred earlier, and secondly, that the whole must have only those properties already present in the parts. The first posed the problem of continuity and discontinuity in evolution, the second that of novel properties in the part-whole relation.

The principle of the continuity of life and its properties had already been defended by Darwin, for whom differences between the lower and the higher orders of life were of degree and not of kind. If a property – such as intelligence or mind – was found at the summit of the evolutionary hierarchy, as in humans, then it must also exist, though less quantitatively developed, at the base in simpler organisms. In a fully developed panpsychist theory, this continuity was extended to the level of the inorganic as well, with the result that intelligence to some degree or other was universal. Lewes ridiculed with irony such a point of view: "By this line of argument one might maintain that not only were there 'sermons in stones', but that these stones were conscious of their eloquence. Pebbles are philosophers of infinitesimal energy." (1879, 2: 31)

But mere ridicule did not suffice, and Lewes addressed the problem theoretically by distinguishing between the law of continuity, which was abstract and quantitative, and the law of discontinuity, which was concrete and qualitative. Though thought was driven to quantitative continuity as an ideal construction by the model of mathematical quantification, this "Law of Continuity" was not a complete comprehension of the world: "That law is simply a deduction from the conception of Quantity, abstracted from Quality by mathematical artifice: it is an abstract idea of Existence irrespective of all concrete Modes of Existence." (1879, 2: 32) To the contrary, Lewes held that discontinuity and qualitative change were present at the level of concrete experience. Speculation was quantitative, and the gaps between things were filled in by a rationalizing intellect:

> If Continuity is a necessity of thought, not less imperiously is Discontinuity a necessity of experience, given in every qualitative difference. ... The manifold of sense is not to be gainsaid by a speculative resolution of all diversities into gradations. Experience

> knows sharply defined differences, which make gaps between things. Speculation may imagine these gaps filled, some unbroken continuity of existence linking all things. (1879, 2: 32-33)

Moreover, like Mill before him, Lewes stated that complexes could have properties that none of their components had. This applied as much to water as a complex of hydrogen and oxygen as it did to life itself, and the complex organic structures upon which it depended. In *Problems of Life and Mind* (first series, 1875) Lewes had already formulated the emergent-resultant distinction. Emergents played a corresponding role in Lewes' system to that of heteropathic laws in Mill's theory:

> Every resultant is either a sum or a difference of the cooperant forces; their sum, when their directions are the same – their difference, when their directions are contrary. Further, every resultant is clearly traceable in its components, because these are homogeneous and commensurable. ... It is otherwise with emergents, when, instead of adding measurable motion to measurable motion, or things of one kind to other individuals of their kind, there is a co-operation of things of unlike kinds. ... The emergent is unlike its components in so far as these are incommensurable, and it cannot be reduced to their sum or their difference. (1875, 2: 413)

In the case of emergents, deduction did not suffice to determine the effect of known causes, and in addition experience was required. Once again, Lewes' examples were from chemistry. Though he mentioned Darwinian evolution approvingly (1877: 45), he did not combine the concept of emergence with that of evolution. Nonetheless, this distinction between emergent and resultant represented a major development in Lewes' thinking from his earlier writings where, influenced by Comte's positivism, he had seen the differences between matter, life, and mind in quantitative terms only.[4]

Lewes further differed from Comte by admitting psychology as a science, one which was based jointly on sociology and biology, but which featured new concepts and laws of its own. Though Lewes recognized that it was a great step forward for science to have recognized the physiological basis of mind, psychology was still one-sided so long as it did not take the further step of recognizing the role of the social in the constitution and development of mind: "Psychology is the analysis and classification of the sentient functions and faculties, revealed to observation and induction, completed by the reduction of them to their conditions of existence, biological and sociological." (1879, 1: 6) Animal psychology concerned itself solely with the relation of the organism and its external, physical medium, but human psychology must add as well the social medium: "In relation to Nature, man is animal; in relation to Culture, he is social." (1879, 1: 71)

But psychology involved more than just biology and psychology, and contained concepts of its own, including that of "faculty," which went beyond the merely biological concept of function. An organic structure had its innate functions – such as the ability of the hand to grasp – which humans developed into faculties, or learned adaptations – such as the ability of the hand to sew: "That is to say, let function stand for the native endowment of the organ, and faculty for its acquired variation of activity." (1879, 1: 27)

The distinction between function and faculty was of importance in the radical break which Lewes believed existed between animals and humans. Though corresponding organs of humans and non-human primates had the same functions, they differed in the faculties to which they were put. It was language as a social product of a special kind which enabled humans to synthesize their experience and go beyond mere functions to faculties. As a result, Lewes explicitly excluded the possibility of animal societies in the same sense as the human. Animal aggregations, be they colonies of bees or groups of apes, lacked the defining characteristics of a true society, which in the following quote appeared rather like Victorian England:

> In the so-called animal societies, there is apparently nothing beyond an aggregation of individuals, with some form of division of employments; there is no subordination nor coordination – only cooperation; no powers invested in individuals and classes; no command and obedience; no relinquishment of personal claims; above all, they have developed nothing like the Family as the social unit, and Tradition as the social experience. (1879: 1: 143)

As a consequence of this qualitative distinction between animals and humans, Lewes criticized Darwin for attributing greater possibilities of mental evolution in animals than could be attained by them. Mind in its fully developed sense must be limited to humans, who alone possessed social organization, language, and the generalized capability of transforming functions into faculties. Mind was an emergent at the human level, a novelty which did not exist in the animal and pre-human world.

Lewes developed a number of themes to be met later in emergent evolution: the combination of continuity and discontinuity, and of quantity and quality in the evolutionary process; certain wholes as having properties not possessed, even implicitly, by their components; the inability of deduction, and the necessity for experience, to determine the nature of emergents; the hierarchy of sciences, with higher ones based on lower ones but going beyond them; and psychology as a science in its own right, though based on both biology and sociology. Lloyd Morgan was candid in his admission of the influence of Mill and Lewes on his theory of emergent evolution:

> The concept of emergence was dealt with (to go no further back) by J. S. Mill in his *Logic* (Bk. III, ch. vi, #2) under the discussion of "heteropathic laws" in causation. The word "emergent," as contrasted with "resultant," was suggested by G. H. Lewes in his *Problems of Life and Mind* (Vol. II, Prob. V, ch. iii, p. 412) Both adduce examples from chemistry and from physiology; both deal with properties; both distinguish those properties (a) which are additive and subtractive only, and predictable, from those (b) which are new and unpredictable; both insist on the claim that the latter no less than the former fall under the rubric of universal causation. (1923: 2-3)

However, Lloyd Morgan's statement, though correct, was only partially so. He did not mention the subtle differences between Mill and Lewes which the foregoing analysis has revealed – including the status of psychology as an irreducible science for the latter but not for the former – nor did he mention three other influences on his own work which are just as important, and which the following sections analyze: Henri Bergson, W. T. Marvin, and E. G. Spaulding.

(c) Creative Evolution: Henri Bergson

Henri Bergson (1859-1941), in *Creative Evolution* (1907) argued that novelty was essential to the evolutionary process. This was a new theory of evolution, and one upon which Lloyd Morgan commented extensively in his writings and notes of 1912-1915. Whereas most previous philosophies of evolution had been monistic, including those of Spencer, Romanes, and Lloyd Morgan himself, Bergson, like Wallace, operated within in a dualistic framework. Wallace, however, had appealed to a philosophically dubious "spirit world," and this had diminished his credibility. Bergson, in *Time and Free Will* (1889) and *Matter and Memory* (1896), situated himself in the dualistic tradition of Descartes, which was metaphysical rather than spiritist.

However, Bergson wanted to avoid the major problem inherent in Descartes' treatment of the mind-matter relation – the difficulty of explaining how non-extended, thinking mind interacted with extended, non-thinking matter. Unlike Descartes, who postulated a physical site of interaction – the pineal gland – Bergson proceeded through the conceptual analysis of the relation between memory and perception. Perception was a function of the material brain, while memory, or more accurately, recollection, was a faculty of immaterial mind. Bergson argued that all real perceptions occurred over time, extending from a past into a present. Memory must be called in to bind together this duration of perception, and it was here that Bergson found the essence of mind-body interaction:

> For after having successively studied pure perception and pure memory, we still have to bring them together. If pure recollection is already spirit, and if pure perception is still matter, we ought to be able, by placing ourselves at their meeting place, to throw some light on the reciprocal action of spirit and matter. "Pure" that is to say instantaneous, perception is, in fact, only an ideal, an extreme. Every perception fills a certain depth of duration, prolongs the past into the present, and thereby partakes of memory. So that if we take perception in its concrete form, as a synthesis of pure memory and pure perception, that is to say of mind and matter, we compress within its limits the problem of the union of soul and body. (1911: 325)

Concrete perception was Bergson's equivalent of Descartes' pineal gland, the meeting place for the distinct substances of mind and brain. But Bergson's view of mind-brain dualism did not feature a one-to-one correspondence between mental acts and brain processes, with the attendant equivalence of complexity between the two. Mind, though linked to the brain, transcended the latter in a sense opposed to theories of simple parallelism between mental and cerebral states.[5] Fond of metaphors, Bergson at one point compared the mind to a knife, and the brain to its edge, and elsewhere stated that the mind was to the brain as a picture was to its frame. According to Bergson, the brain was concerned only with the translation of thoughts into action; it was concerned only with space, which was inferior and subordinate to time, the domain of mind.

Related to these distinctions between mind and matter, and time and space, was that between quality and quantity. Quality was a characteristic of mental events in time while quantity was a characteristic of material things in space. Since interaction between matter and mind was permitted, quantity in space could cause or bring about qualities in the mind. But in general, given the priority of mind over matter which Bergson advocated, it was quality which took precedence over quantity.

Just as he argued for a non-material mind responsible for thought, Bergson also believed in a non-material factor responsible for evolution. In *Creative Evolution* he examined as possible modes of change Darwin's gradualism, Hugo de Vries' mutationism, T. H. Eimer's orthogenesis, Lamarck's use-inheritance, and Spencer's transformation of the homogeneous into the heterogeneous. But he found none able to properly explain phenomena such as the development of complex organs in the course of evolution, or convergent evolution through which similar organs developed in unrelated species. Bergson claimed that no natural process could produce these phenomena, and that only a non-material "*élan vital*" – vital force – responsible for life could do so:

> So we come back, by a somewhat roundabout way, to the idea we started from, that an original impetus [*élan vital*] of life, passing from one generation of germs to the following generation of germs through the developed organisms which bridge the interval between the generations. This impetus, sustained right along the lines of evolution among which it gets divided, is the fundamental cause of variations, at least of those that are regularly passed on, that accumulate and create new species. (1911: 87)

The vital spirit confronted matter as its opposite. Its freedom was limited by the fact that it must act through matter, which represented determinism and necessity. The creativity, freedom or indeterminism of evolution manifested itself in the novelty introduced into organic life, including the splitting of evolution into plant, animal, and human lines of development, of which the last was qualitatively higher than the others because of the presence of self-reflective consciousness. Bergson situated his creative evolution as an alternative to mechanism. Mechanism had a limited validity, and was applicable to "systems that our thought artificially detaches from the whole." (1911: 36) But once attention was turned to the whole, to time, and most importantly, to duration, Bergson concluded that mechanism could not capture the fluidity and the uniqueness of life and mind. Creative evolution, by introducing qualitative novelty into the world broke with the mechanistic worldview in a decisive way.

In his analysis of the faculties of mind, Bergson sharply limited the role of the intellect to the analysis of wholes into parts. Such a faculty of analysis was clearly justified, even necessary in order to comprehend inert matter, but it did not suffice to capture the meaning of dynamic processes such as life, mind, and evolution. Instinct, in the developed form that it was found in humans – as intuition – captured the essence of these phenomena through an act of sympathy and self-reflection which intellect could not achieve:

> Instinct is sympathy. If this sympathy could extend its object and also reflect upon itself, it would give us the key to vital operations – just as intelligence, developed and disciplined, guides us into matter. ... But it is to the very inwardness of life that intuition leads us – by intuition I mean instinct that has become disinterested, self-conscious, capable of reflecting upon its object and of enlarging it indefinitely. (1911: 176)

These two claims – the existence of a non-material vital spirit, and the inability of intellect to comprehend life, mind, and evolution – were essential elements of Bergson's thought. However, equally important was his commitment to creativity and novelty in the evolutionary process: "Every human work in which there is invention, every voluntary act in which there is freedom, every movement of an organism that manifests spontaneity, brings something new into the world." (1911: 239) In Bergson's dualistic system, creativity was the basic category within the spiritual realm, and novelty was its phenomenal result within the world of matter. It was this concept of creative evolution productive of novelty which was of interest to Lloyd Morgan. Bergson's system was a subject of considerable debate in England. Some, including Ray Lankester and other Huxleyans, rejected the theory as irrationalist.[6] Others, including Lloyd Morgan, were interested by the concept of qualitative novelty it introduced into the discussion of philosophy of evolution.[7]

In "The Case for Emergent Evolution" (1929), Lloyd Morgan indicated that the first time he used the term "emergent evolution" was in a 1915 article in the review *Scientia*, though he had added the following interleaved note to *Instinct and Experience* shortly after its publication in 1912, indicating the influence of Bergson on his thinking about evolution at that time:

> Why not work things out under heading Emergent Evolution? It would enable one to emphasize a policy of interpretation based on empirical facts of observation, leaving the question: To what may these facts be due? to be discussed as a wholly different issue, e.g. under some doctrine of Creative Evolution" (1929d: 29)

The influence of Bergson on Lloyd Morgan, however, should not be exaggerated. Rudolf Metz in his *A Hundred Years of British Philosophy* (1938), written at a time when emergent evolution was fresh in memory, stated that emergent evolution was just an English form or variant of Bergson's creative evolution:

> In substance, this idea is, of course, not new. What is new is the philosophical range which Alexander and Morgan have given it. ... But it was through Bergson's idea of creative evolution that the doctrine of novelty becomes widely known and made its way into England, where by a similar reaction against the mechanistic evolution theory, Alexander and Morgan became its most influential champions. Emergent evolution is a new, important and specifically British variation of Bergson's creative evolution. (1938: 656)

A number of considerations were left out of Metz' analysis: firstly, the theses of Bergson's theory which Lloyd Morgan specifically rejected, secondly, the modifications which Lloyd Morgan made to those these which he accepted, and thirdly, the other influences on Lloyd Morgan. This will be further examined in chapter eight.

(d) Novel Properties of Wholes: E. G. Spaulding

A final influence on Lloyd Morgan's thought in the crucial period of the beginning of the 1910s was the work of two of the American New Realists, Edward Spaulding and Walter Marvin.[8] Spaulding (1873-1940), in "A Defense of Analysis" (1912), rejected the limitations Bergson placed on analysis, but at the same time he developed a part-whole theory essential for emergentism. He argued that not only was analysis legitimate as an intellectual activity in all domains, but that things, when analyzed, could in some cases be shown to have properties that none of their parts possessed:

> Given a whole which, for one reason or another, is known to be analyzable, then analysis reveals parts, but it also reveals the relations which relate and so organize these parts into some kind of a whole. Consider also those properties which, in some cases, the whole, as a whole, may have different from those of the parts. Of course, analysis reveals these also. (1912: 161)

Though analysis might be incomplete, in the sense that the parts could be analyzed into further parts, analysis did provide an adequate or valid decomposition of the whole into its immediate constituents, their properties, and their relations. Spaulding distinguished between various types of wholes which could be the object of analysis. The first kind was the aggregate or collection, where the basic relation uniting the parts was that of simple conjunction. Such a whole might consist of disparate parts, such as a heap of different objects, and involved no novelty beyond that which was already present in its parts. Other types of wholes were those made of individuals which were encompassed under a common universal, and those which were compounded out of subordinate classes. But the highest type of whole was the "organic whole," by which term Spaulding referred to both chemical compounds and living organisms, and it was here that novel properties appeared:

> If the whole be experimentally synthesized out of the parts, then something new appears as properties of the whole, something which is new qualitatively as well as quantitatively. On the other hand, if an experimental analysis be made of the whole, then the whole is also found to have properties which the parts do not have. These properties are put "in relief" by the analysis; they are a residuum, characteristic of the whole as a whole, and revealed by analysis which at the same time reveals the parts or elements, and through its ramifications, the organizing relations. (1912: 237-38)

In developing this idea, Spaulding referred to the work of the chemist Walter Nernst in *Theoretical Chemistry from the Standpoint of Avogadro's Rule and Thermodynamics* (1895). Nernst made the distinction between an additive property, where the value of a property of the molecule was the sum of the values of the properties of its component atoms, and constitutive properties, where this was not the case. Nernst based his distinction on the work of Wilhelm Ostwald.[9] Spaulding's formulation of the distinction (1912: 238), which included quotes from Nernst, was reproduced by Lloyd Morgan in footnote 54 of *Spencer's Philosophy of Evolution* as follows:

> This fact, that in the actual synthesis, artificial or natural and developmental, of existential wholes out of parts, new properties or values appear, is a matter of great importance. It is a fact, too, which is accepted by authoritative investigators. Says Professor Nernst, "A large number of physical properties have been shown to be clearly additive; that is, the value of the property in question can be calculated as though the compound were such a mixture of its elements that they experience no change in their properties." Examples are the volume, refraction,

> magnetism, and heat of combustion of organic compounds. But other properties are not additive. "The kind of influence of the atom in a compound is primarily dependent upon the mode of its union, that is, upon the constitution and configuration of the compound. Such non-additive properties are called constitutive." Examples are the absorption of light, the rotatory power, the melting point. (1913: 50)

There was, however, a difference between Nernst's concept and Spaulding's interpretation of it which was significant for the subsequent development of emergent evolution. In discussing additive and constitutive properties, Nernst was not explicitly dealing with a new property of the molecule different from any of the properties of the constituent atoms. Rather, the numerical value of a property common to the molecule and its constituent atoms was such that the former was not a simple arithmetic sum of the latter. Spaulding injected an interpretation of his own in going from quantitatively non-additive properties, to qualitatively novel properties, and it was this aspect which struck Lloyd Morgan as most significant.

Spaulding went on to develop his proto-emergentist ideas in *The New Rationalism* (1918).[10] His acceptance of whole-part emergence was clearly expressed as follows: "In the physical world (and elsewhere) it is an established empirical fact, that parts as non-additively organized form a whole which has characteristics that are qualitatively different from the characteristics of the parts." (1918: 447) He referred to this process whereby novel qualities appeared in the evolutionary process as "creative synthesis."[11] He also admitted a level structure of reality, and distinguished between causal and functional relationships. Causal relationships held within a level, while functional relations held between higher and lower levels. Such related levels were said to be compatible with one another, but not reducible one to the other. Spaulding mentioned among the irreducible but compatible levels of reality the physico-chemical, the biological, and the ethical, though he did not develop on this structure in greater detail.

(e) Logical Strata of Reality: Walter T. Marvin

The second of the American New Realists who influenced Lloyd Morgan was Walter T. Marvin (1872-1944), who set out his basic views on the logical strata of reality and argued for a moderate form of creative evolution in *A First Book in Metaphysics* (1912). The book, as its title suggested, was intended as an introduction to philosophy, or as Marvin put it "a student's first book." It was therefore surprising that Lloyd Morgan, who certainly could not be considered a beginner in philosophy, should refer to it. But the volume contained more than a list of philosophic doctrines of the past, including as it did a sophisticated discussion of the concept of the logical strata of reality.

Unlike Spaulding, Marvin aligned himself with Bergson on the problems of time and creative evolution. Marvin contrasted temporalism, which treated time as a fundamental aspect of reality, with eternalism which did not. Within temporalism, he distinguished between causal pluralist theories which admitted the unique and novel in the course of temporal evolution, and causal monist theories which did not. Marvin argued for temporalism of the causal pluralist type, and in his references for further study on time asymmetry, recommended two books by Bergson – *Time and Free Will* and *Creative Evolution*. The filiation of ideas was most clearly revealed when Marvin stated his belief that history was not a mere repetition of the events of the past, at most in new combinations, but involved novel events that could not be so analyzed:

> New things, new events, and perhaps even new elementary constituents of these things and events are potentially in the present. This fundamental newness, and the resulting impossibility of knowing the future fully may be called freedom or spontaneity, or better, creative evolution. Rigorously defined, creative evolution is the truth that some particular existential propositions related to future instants are not so related to past instants and that these propositions are neither deducible from universals or from the past, nor knowable beforehand in any other way. (1912: 135)

But unlike Bergson, who grounded creativity in the *élan vital*, Marvin found its source in chance and particularity. He argued that individuals were more than mere particulars falling under universals. They were also centers of spontaneity and creativity, subject not only to causal laws, but to chance occurrences as well. Science aimed at discovering general laws, but was limited both by their potentially infinite number and by the ultimate uniqueness of individuals. What science did reveal was that reality could be analyzed as a series of "logical strata", where each successive stratum included fewer but more specialized types of entities. There was a tension between the mind's desire for the continuity of these logical strata, and the discontinuities introduced by the spontaneity of the "evolving existent." Though it was an idealization to analyze reality into logical strata, it was one which science was justified in making, so long as it did not forget the limitations of this act imposed by the individuals which made up the world. Marvin identified the series of strata as follows:

> To sum up: the picture of reality just outlined is logically built up of strata. The logical and mathematical are fundamental and universal. The physical comes next and though less extensive is still practically, if not quite, universal. Next comes the biological, extensive but vastly less extensive than the chemical. Finally, comes the mental and especially the human and the social, far less extensive. (1912: 143-44)

Marvin noted that each of these logical strata could itself be analyzed into substrata so that the listing given above, while considered sufficient, might not be definitive. The student of a higher level must be familiar with the lower levels, upon which the higher one depended, but the lower levels were independent of the higher ones which subsequently arose from them. Marvin's concept of the logical strata of reality completed the further influences on Lloyd Morgan in his search for a theory of evolutionary novelty.

CHAPTER EIGHT

Lloyd Morgan's Creative Synthesis

Lloyd Morgan now had before him the essential components out of which his own "creative synthesis" of emergent evolution would develop: the evolutionary theories of Darwin, Spencer, Huxley, Wallace, and Romanes, Bergson's notion of creative evolution productive of qualitative novelty, the theories of heteropathic laws due to Mill, and emergent properties as formulated by Lewes, Spaulding's concept of novelty in the whole-part relationship, and the theory of levels of reality developed by Marvin. In the following section, I will argue that Lloyd Morgan analyzed these background theories, selected and rejected their theses, transformed the retained theses and then integrated them into a new theory which included an innovative element – the placing of emergence at the heart of a monistic philosophy of evolution. This was the project that Lloyd Morgan accomplished in the period from 1912 to 1915.

(a) Lloyd Morgan on Bergson

Lloyd Morgan first examined Bergson's views in *Instinct and Experience* (1912), and corresponded with the French philosopher on this subject.[1] Though he disagreed with Bergson's view of instinct as superior to intellect, Lloyd Morgan agreed with Bergson that evolution was productive of novelty, and affirmed that there were two factors present in evolution: "(1) some measure of substantial but never complete repetition, and (2) some measure of the new and unique." (1912a: 171) His attitude to novelty was still ambivalent, and he left open the possibility that the new might simply be an algebraic – though not arithmetic – function of its components, and therefore derivable from the component parts given their magnitudes and the algebraic function into which they entered as arguments:

> Here again, however, we are faced with the same difficulty of interpretation. Is the apparently new and unique a veritable "creative" departure from routine? Or is it the algebraical sum of characters given in previous routines and therefore predictable if we know the amounts of these characters and the mode of their summation? I see, at present, no ground for denying, though I am not prepared to assert, that really new synthetic combinations, as contrasted with quasi-mechanical mixtures of old characters, do occur in the natural history of experience. But since as matters now are, we have not the data for proof of either

their presence or absence, let us be content to grant that they may occur. (1912a: 173)

It was in the crucial year of 1912 that Lloyd Morgan completed the intellectual journey that resulted in the theory of emergent evolution. On Feb. 24, 1912, he wrote out four pages of notes under the title "Creative Evolution" for a talk which he gave to a Saturday night audience at the University of Bristol, and again, on March 15, 1912, he wrote out eleven pages of notes entitled "Some aspects of M. Bergson's Philosophy." He began his February notes with a threefold summary of the position of Bergson, which he repeated in the March notes as well:

Now Mr. Bergson's main points are these:
1. Evolution is history of progressive change.
2. History does not repeat itself.
3. In the absence of a routine of repetition we cannot foresee what is coming.
4. The unforeseeable in evolution is creative.

Again,
1. Evolution is not only a succession of products but a continuous process.
2. Any stage of organic process embodies the net results of all previous stages of process.
3. But process grows as it flows onwards.
4. The new elements of growth which go beyond previous routine are creative.

Once more
1. Any process involves a Source or Cause.
2. The Source or Cause of organic process is Life itself (*l'élan vital*).
X 3. Life makes organic machines for its use and when so made they go by routine.
4. But beyond mechanical routine there are free developments. These constitute creative evolution. (1912b: 1)

There was an important difference between the first two and the third of the above summaries. The first two formulations argued for creativity and novelty in the evolutionary process, and Lloyd Morgan would incorporate them in his own theory of emergent evolution. But the third summary concerned Bergson's specific claim that the Source or Cause of creativity was the *élan vital*, which Lloyd Morgan rejected since he reserved Source or Cause for Deity. He indicated his disagreement with this way of looking at evolution by an "X" to the left of the third argument, next to statement three, but referring to the whole third argument. In his further commentary Lloyd Morgan noted that "In science do not consider Source. Process and products, conditions and constitution suffice." (1912b: 2) He concluded his critique of Bergson with the comment: "But Mr. Bergson says that Life creates, that Life is the Source of organization. That I contend is not a scientific proposition. The question what is the Source of phenomena is a great question, perhaps *the* great *question*. But it does not admit of a *scientific* answer." (1912b: 4, italicized words were underlined in the original) The answer was ultimately philosophical and theological: as Lloyd Morgan had argued in his work of the 1890s, the Source was Deity.

But the first two formulations as given above, which did not refer to source or cause, were much more attractive to Lloyd Morgan, so much so that he eventually, and within the year, adopted them. He asked himself, "What does M. Bergson mean by saying that history does not repeat itself? Let us pause to consider: If the same tune of life is played over and over again *da capo* how can there be Evolution?" (1912b: 3) He then went on to consider a variety of examples, from the origin of horses, and the wings of vampire bats, to human history, and asked himself whether old routines could be combined in new ways, and if so, would there be continuity or saltation? Lloyd Morgan began to accept the notion of unpredictability in novel events: "Explanation always after the event. No one can predict what the universe has in store for it. New routines may come in. We live forwards and explain backwards." (1912b: 4) At the cosmic level, unpredictable novelties occurred. The stage was set for emergent evolution.

(b) Lectures on Scientific Thought

The final development of the concept of emergent was contained in Lloyd Morgan's lecture notes for a course on "Scientific Thought" prepared in the fall of 1912, probably after mid-October.[2] As indicated above, he was considering Bergson's theory of creative evolution at the time. Bergson's philosophy of creative evolution and irreducible novelty was a necessary, but not a sufficient stimulus for Lloyd Morgan's work on emergent evolution, and a further influence from chemistry was required. Lloyd Morgan returned full circle to his earliest work on water and other chemical phenomena, but his analysis differed substantially from that of thirty years previous.

In speaking of the changes of state of water in melting or boiling, he noted that there was "a sudden *spring* from one molecular state of the physical substance to another. The one does not *slide* or glide into another." (1912d, Archives 128/553: 104, italics were underlined in the text) And he continued as follows concerning the more significant chemical change when hydrogen and oxygen combine to form water: "The old adage *Natura nihil facit per saltum* must not be taken too literally. When oxygen and hydrogen combine to form water a new substance with new properties suddenly springs into being." (1912d, Archives 128/553: 104) This was a major shift from his own 1885 work on water, where he expressly denied this "sudden springing into being" of new properties, and like Darwin supported the "old adage" that nature made no leaps.

It was in the section entitled "Additive and Constitutive Properties" that the pieces of the puzzle fell into place. Lloyd Morgan referred to Mill's *Logic* and his discussion of composition of causes and heteropathic laws. Lloyd Morgan then associated Mill's terminology with that of Nernst as interpreted by Spaulding: composition of causes corresponded to additive properties, while heteropathic laws gave rise to constitutive properties. Quoting Mill on "heteropathic laws" from Bk. iii, Ch. vi of his *Logic*, Lloyd Morgan then added as a marginal note: "cf Resultants and Emergents – Lewes Problems Vol. ii. p. 412." (1912d, Archives 128/553, p. 117) It was at this moment that Lloyd Morgan put three further elements of emergentism together – Mill's heteropathic laws, Lewes' terminology of emergents, and Spaulding's gloss on Nernst's constitutive properties.

(c) Lloyd Morgan's Spencer Lecture

Whereas in *Instinct and Experience* Lloyd Morgan's tone was still hypothetical, in "Spencer's Philosophy of Evolution," delivered as the Spencer Lecture for 1913, he shifted to a categorical position on evolutionary novelty and the explicit use of emergentist terminology. The background to this was his reading of Bergson's work on evolution, but the foreground, the immediate context, was Lloyd Morgan's re-reading of the problem in chemistry of relating the properties of water as a compound to those of its atomic components. All the basic elements for the development of the core concept of emergent evolution were present in this 1913 text, and it was therefore the crucial public text in the process of creative synthesis that resulted in the theory of emergent evolution.

Lloyd Morgan's Spencer Lecture was devoted to an exposition and evaluation of Spencer's philosophy of evolution – a Spencer lecture on Spencer. It also included Lloyd Morgan's evaluation of Bergson's criticism of Spencer. Bergson claimed that Spencer reduced the concept of evolution to that of mechanical action.

Lloyd Morgan disagreed in one respect, since he considered that Spencer's Unknowable, similar to his own Source, was the metaphysical cause of evolution, distinct from and conditioning it, and not a mechanism or mechanical agent. But in another respect, Lloyd Morgan agreed with Bergson, since he considered that Spencer had adopted a mechanistic formulation in analyzing the content of evolution. In Spencer's own terms, evolution produced the successive domains of the inorganic, organic, and super-organic. Lloyd Morgan's criticism was that Spencer did not sufficiently explain the qualitative diversity of these domains, or as Lloyd Morgan now put it, the different types of "relatedness" involved at each stage. By arguing that the single process of differentiation of the homogeneous into the heterogeneous was the common source of these three domains, Spencer had reduced the different kinds of relatedness – though not evolution itself – to the mechanical type.

Lloyd Morgan argued that new kinds of relatedness appeared in nature and evolution. The preformationist process of making explicit that which was implicit occurred only at the level of metaphysical evolution, where deity was the source or cause of evolution. Physical evolution was an epigenetic process characterized by the appearance of novelty. Bergson's term "creativity," shorn of its dualistic connotations, was applied to the metaphysical source and therefore reserved for deity, while "novelty" was the term used to describe the appearance of new phenomena in the evolutionary process as studied by science.

With this perspective in mind, Lloyd Morgan proceeded to the first formulation of his matter, life, and mind theory of the levels of reality. There were three main types of relatedness to be examined, each symbolized by a letter: *A* for relations of the physico-chemical type, *B* for relations of the organic type, and *C* for those of the cognitive type. Lloyd Morgan said of the level of matter: "Now the successive products, in which this physico-chemical type of relatedness obtains, have certain new and distinctive properties which are not merely the algebraic sum of the properties of the component things prior to synthesis." (1913b: 28) The reference was to Spaulding's interpretation of Nernst's distinction between constitutive and additive properties, as discussed in the previous section. The earlier suggestion that novel properties might just be "algebraic" functions of the component properties, as expressed the previous year in *Instinct and Intelligence*, was now abandoned.

Lloyd Morgan adopted the concept of a hierarchy of evolutionary stages, with each higher stage grounded in the previous stage, and with each new stage characterized by the emergence of a new type of relation. This proposition was footnoted to Marvin's discussion of the strata or levels of reality: "For just as higher types of relatedness imply a substratum of physico-chemical processes, so do all events imply the underlying logic of events. Cf. W. T. Marvin, *A First Book in Metaphysics*, ch. xiii, 'On the Logical Strata of Reality'." (1913: 51-52, fn. 72) The confluence of influences was now complete.

(d) Lloyd Morgan's Creative Synthesis

I designate the process of the formulation of a new theory by a critical reading of previous theories as "creative synthesis," borrowing this term from Spaulding. New theories do not come out of the void; they are based on antecedent theories. Lloyd Morgan based himself on previous thinkers, accepted certain of their theses, rejected others, modified the accepted theses, and combined them in an innovative way. The borrowed elements provided a continuity upon which the qualitative novelty of innovation arose. The innovation was therefore relative and not absolute.

A comparison can be made with the way in which Darwin developed his theory of species evolution through natural selection. Contrary to T. H. Kuhn's claim in *The Structure of Scientific Revolutions* (1969) that theories before and after a scientific revolution are incommensurable, there was a degree of continuity between Darwin and his predecessors – including Darwin's philosophical debt to Comte's theory of the stages of knowledge, his acceptance of the principle of continuity expressed by Leibniz as "*natura non facit saltum*," the influence of Malthus' population theory, and the further influence of Milne-Edwards' concept of the biological division of labour. Darwin critically analyzed the theories of his predecessors, accepted, modified, and rejected various elements, in order to produce his own theory of evolution by natural selection. This does not mean that his theory were merely a combination of old ideas in modified guise. The genuine novelty in Darwin's theory was the innovative thesis of natural selection on analogy to artificial selection.

This model of creative synthesis can be applied to Lloyd Morgan as well, though with a significant qualification. As described above, Lloyd Morgan performed a critical analysis of the theories of Bergson, Mill, Lewes, Spaulding, and Marvin, as well as those of Darwin, Spencer, Huxley, Wallace, and Romanes, and selected and rejected relevant ideas for his theory of the evolution of novelty. This was the first aspect – that of critical analysis – confirming that a creative synthesis was under way. The second aspect, the innovative element, was more problematic and requires that a distinction be made between a strong and weak sense of innovation.

The analysis of Lloyd Morgan's emergentist writings does not reveal an innovative thesis comparable in nature and originality to Darwin's thesis of natural selection. The concept and indeed the term "emergent" had already been developed by Lewes following Mill, and Lloyd Morgan himself recognized this. What was innovative was the role Lloyd Morgan accorded to the concept of emergence. Unlike Mill, for whom heteropathic laws were subsidiary to laws of composition, and Lewes, who did not combine evolution and emergence in a unified theory, Lloyd Morgan placed emergence at the center of his philosophy of evolution. In contrast to Darwin's development of an innovative thesis, Lloyd Morgan's innovation was one of emphasis and role for a borrowed thesis, a creative synthesis of a structural rather than a substantial type.

CHAPTER NINE

Lloyd Morgan's System of Emergent Evolution

Lloyd Morgan developed his system of emergent evolution in the two volumes of his Gifford Lectures, delivered in 1922-23 and published as *Emergent Evolution* (1923) and *Life, Spirit and Mind* (1926). He also set out his philosophy in the last of his books, *Emergence of Novelty* (1933), and in articles published in the *Monist, Journal of Philosophy, Philosophical Studies*, and other reviews. The theses of his developed system can be divided into four groups: those concerning emergence and evolution, those concerning the levels of reality, those concerning the nature of mind, and those concerning the evolution of society. While Lloyd Morgan made a considerable contribution to the philosophical analysis of novelty in evolution, his work was not without serious problems which resulted from his combination of emergentist concepts with dual aspect notions related to neutral monism.

(a) Emergence and Evolution

Central to Lloyd Morgan's concept of emergent evolution was the thesis of the co-occurrence of emergents and resultants. Resultants could occur without any accompanying emergents, but all emergents occurred in the context of at least some resultant effects. Emergents were not saltationist breaks, but instead marked changes in direction in the continuous path of evolution. They were not stray properties, isolated from all others, but were associated with the resultant effects which accompanied them, formed their basis, and provided an underlying quantitative continuity for them:

> Resultants give quantitative continuity which underlies new constitutive steps in emergence. And the emergent step, thought it may seem more or less saltatory, is best regarded as a qualitative change of direction, or critical turning point, in the course of events. In that sense there is not the discontinuous break of a gap or hiatus. It may be said, then, that through resultants there is continuity in progress; through emergence there is progress in continuity. (1923: 5)

It was essential for Lloyd Morgan to incorporate an element of continuity in emergent evolution, in order to avoid a saltationist theory. Yet the anti-saltationist thesis *natura non facit saltum* – nature makes no leaps – had been an obstacle to the development of emergent evolution, since it had been used to cover both the continuity of evolution, and the exclusively quantitative nature of evolutionary change. Lloyd Morgan disconnected the two aspects of Darwin's thesis, and though he agreed that evolution was a continuous process, he admitted qualitative novelty as well.

Lloyd Morgan held that it was not possible in advance of direct experience to predict the character of events emergent at a higher level: "What, it is claimed, one cannot predict, then, is the emergent expression of some new kind of relatedness among pre-existent events." (1923: 6) But the predictability of knowledge was an epistemological consideration, while the status of novelties as reducible or emergent was ontological. The link between these two propositions was by no means as obvious as Lloyd Morgan believed. The inability to predict emergents might be a characteristic of the current state of science, to be overcome at a later stage of development, a possibility which Lloyd Morgan did not consider.

Lloyd Morgan used the term "supervenience" to denote the return action of emergents upon the lower level events from which they arose. Events at a higher level "involved" associated events at lower levels, and were based on them. At the same time, the higher level events influenced the lower level ones from which they arose, which therefore "depended" upon them and took place differently than in their absence:

> I speak of events at any given level in the pyramid of emergent evolution as "involving" concurrent events at lower levels. ... But when some new kind of relatedness is supervenient (say at the level of life), the way in which the physical events which are involved run their course is different in virtue of its presence – different from what it would have been if life had been absent. ... I shall say that this new manner in which lower events happen – this touch of novelty in evolutionary advance – depends on the new kind of relatedness." (1923: 15-16)

If emergents were not to be mere epiphenomena, they must have a return causal role – termed "downward causation" by some later emergentists – and the thesis of supervenience was an acknowledgement of just such a role. But Lloyd Morgan was not explicit whether the "touch of novelty" induced on lower level events by higher level ones was qualitative, in which case higher level laws would modify lower level laws, or quantitative, in the sense that only the rate, distribution, and extent of lower level processes were affected by the higher level ones.

Although evolution most commonly proceeded from lower to higher levels and from less complex to more complex entities, it could also bring about degradation or loss of properties and qualities. Lloyd Morgan admitted both chemical evolution from less complex hydrogen to more complex uranium, and the reverse process or "devolution" of uranium into simpler, non-radioactive byproducts through fission. But given that he believed that the course of evolution, under the stimulus of its source or agency, was essentially productive of novel types of emergent reality, he committed himself to the thesis of the progressive nature of evolution – the predominance of evolution over devolution.

Lloyd Morgan considered as scientific in nature the thesis that evolution resulted in qualitative novelty, the "outspringing of something that has hitherto not been in being," (1923: 112). However, emergent evolution did not propose a specific law or factor of evolution for biology, or for any other domain of reality studied by a special science. Rather, it set out the philosophical claim that there was both continuity and novelty in the transition from matter to life, and again in the transition from life to mind, in opposition to reductionist, saltationist, and supernaturalist interpretations. Emergent evolution related the domains studied by the sciences of physics, chemistry, biology, and psychology – a philosophical task not undertaken by any one of them – but did not propose mechanisms of change specific to any one of them – a scientific task which philosophy could not undertake.

The consequence of Lloyd Morgan's misapprehension of the status of emergent evolution was that he advanced a second theory of evolution as a philosophical interpretation. In the philosophical sense, he believed that evolution was "the unfolding of that which is enfolded; the rendering explicit of that which was hitherto implicit." (1923: 111) This may be termed "metaphysical evolution" and invoked not the epigenetic process typical of emergentism, but rather a preformationist process linked to teleology, with the end or goal pre-determined from the beginning. This, however, weakened the argument for emergence as an alternative to both mechanism and teleology, and like Bergson's creative evolution, skewed philosophy of evolution unilaterally towards the latter. Moreover, the philosophical aspect of evolution, which did not involve novelty, was now at variance with its scientific aspect, which did.

Lloyd Morgan was led into this contradiction because of his view that science should be harmonized with theology. Given that nothing was novel for the omniscient and omnipotent deity, metaphysical evolution as the handiwork of the divine had to feature preformationist rather than epigenetic change. To complete this reconciliation of evolution with theology, Lloyd Morgan adopted the thesis of "*nisus* towards deity" – a phrase borrowed from his colleague Samuel Alexander.

Deity was not only the source or agency behind evolution, but the end or goal towards which it tended. The acceptance of Deity as both source and end was recognized with "natural piety" – the belief that the fact of emergence could not be further analyzed by science or a scientifically-based philosophy, and could be explained only through an appeal to theology and a theologically-based metaphysics. Like the natural theologians who studied natural history to discover the wisdom and goodness of God, Lloyd Morgan saw in emergent evolution the product of a creative divinity:

> Under naturalistic treatment, however, the emergence, in all its ascending grades, is loyally accepted, on the evidence, with natural piety. That it cannot be mechanically interpreted in terms of resultants only, is just that for which it is our aim to contend with reiterated emphasis. But that it can only be explained by invoking some chemical force, some vital élan, some entelechy, in some sense extra-natural, appears to us to be questionable metaphysics. It may be that we have just to accept the newly given facts – all the facts as we find them – in the frankly agnostic attitude proper to science. Or it may be that in the acknowledgement of God an ultimate philosophical explanation, supplementary to scientific interpretation, is to be found. That will be the position I shall try to maintain. (1923: 8-9)

The principles of *nisus* towards deity and natural piety weakened the influence of emergentism as a trend of philosophical thought by tying it to theological commitments, and motivated the rejection of emergent evolution by Bertrand Russell and some other commentators of the 1920s and 1930s. Lloyd Morgan's system was based on the core principles of evolution as productive of qualitative novelty, and the emergence of distinct and irreducible levels of reality. By including principles related to theology, Lloyd Morgan multiplied the number of assumptions beyond those which were needed for a philosophy based on evolution, a process which, as an object of science, does not require deity as its source or goal.

In conclusion, the following were the basic theses of Lloyd Morgan's theory of emergent evolution. The title for each thesis was suggested by his discussion, though not always explicitly so named by him:

Thesis of the co-occurrence of emergences and resultants: Emergents are not stray events, but always occur along with resultants.

Thesis of quantitative continuity and qualitative novelty: The resultants provide the quantitative continuity upon which qualitative novelties arise from time to time as changes in the direction of evolution, and not as saltationist leaps.

Thesis of the supervenience of emergents: Emergent properties in systems arise upon the basis of lower level ones, but there is also a return action, or supervenience of the former on the latter, so that the lower level processes proceed "differently"– in a sense not further analyzed – under the influence of the higher level emergents.

Thesis of unpredictability of emergents: Emergents are unpredictable even when given complete knowledge of the antecedent events, component parts, and applicable laws.

Thesis of progressive evolution: Evolutionary change is progressive, and on balance, evolution along with the emergence of novelty dominates devolution and the loss of qualities.

Thesis of physical epigenesis: Physical evolution, or evolution as studied by science, is a developmental process of evolutionary stages characterized by the appearance of novel phenomena.

Thesis of metaphysical preformationism: Metaphysical evolution is the rendering explicit by a divine source or agency of that which is implicit in matter and its capacities for organization

Thesis of nisus towards deity: Deity is the Source or Activity which underlies evolution, and evolution tends to realize the goals or ends set by this supernatural source.

Thesis of natural piety: Emergence must be accepted with "natural piety"; its mechanism cannot be further analyzed in scientific or purely philosophical terms, but involves additional reference to theology.

(b) Emergence and the Levels of Reality

Lloyd Morgan began his *Emergent Evolution* with a discussion of Alexander's *Space, Time and Deity* (1920), and his presentation of the emergent levels of reality can best be followed by examining the similarities and differences between his schema and that of Alexander, who will be further discussed in chapter ten. Though Lloyd Morgan had formulated his central ideas about emergent evolution in the period 1912-15, he did not publish his major work *Emergent Evolution* until 1923, based on his Gifford Lectures of the preceding year. Influenced by Lloyd Morgan, Samuel Alexander had earlier presented his own emergentist system *Space, Time and Deity* as the Gifford Lectures for 1916-1918. Thus, Lloyd Morgan, the originator of the theory, found himself commenting on a collaborator's previously published work. However, the relation between Lloyd Morgan and Alexander was a cooperative one, and Alexander recognized Lloyd Morgan's priority in the formulation of emergent evolution.

According to Lloyd Morgan, Alexander had argued for a structure of reality characterized by five levels. The basic level was that of space-time, from which emerged matter, with its primary and secondary properties, followed by life and mind; the highest level was that of deity. Lloyd Morgan presented Alexander's analysis in diagrammatic form as follows: (1923: 11)

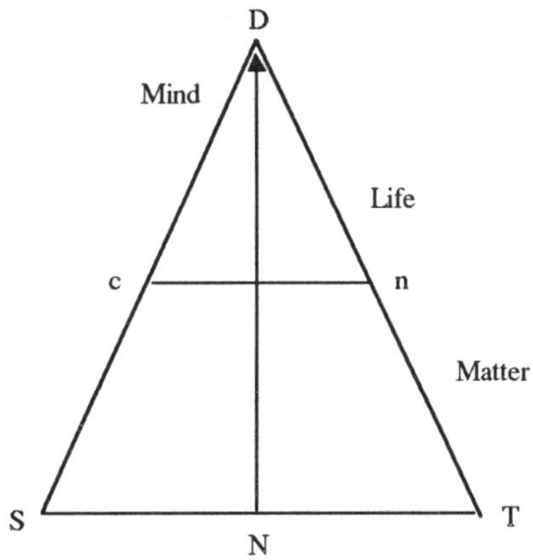

Diagram 9.1: Lloyd Morgan's Representation of Emergent Evolution

In Lloyd Morgan's presentation, the structure was simplified by representing Deity by the point *D* at the apex of the triangle, and Space-Time by the line *ST* at the base. This left in the middle just the three levels of Matter, Life, and Mind which Lloyd Morgan admitted. Matter and Life were thought of as primarily physical, but by the dual aspect thesis of neutral monism, they also had psychical correlates, while Mind, which was thought of as primarily psychical, had a physical correlate: "Hence our comprehensive scheme runs thus: C, Mind (with physical correlates); B, Life (with psychical correlates); A, Matter (with psychical correlates)." (1923: 27) The *ND* arrow represented the "*nisus* towards deity" which was the driving force of evolutionary advance, corresponding to what Lloyd Morgan had earlier termed the Source or Agency underlying evolution. A slightly different diagram of emergent evolution was given in Lloyd Morgan's lecture notes for his Huxley Lecture, delivered in Birmingham, Nov. 25, 1921:

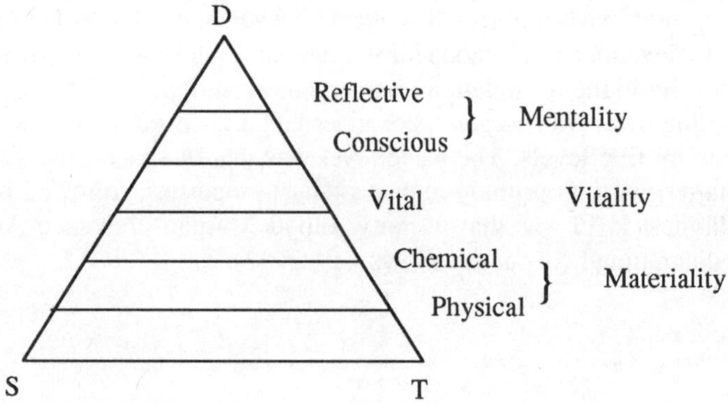

Diagram 9.2: Lloyd Morgan's 1921 "Pyramid of Levels"

In his notes, Lloyd Morgan remarked: "Chief speculative difficulties [are] at base and at apex" (1921c: 4) At the base, he continued, were "what we may call with some other thinkers neutral entities." (1921: 4) These he identified as Alexander's pieces of space-time with their primitive motions, or Alfred North Whitehead's "event-particulars." At the apex was God or Deity. But Lloyd Morgan observed that this further speculation might not be needed in the theory of emergent evolution itself:

> Shall we start with the physical plane and not at present go beyond the reflective plane? That, I think, would have been Huxley's alternative. And it leaves enough on our hands. ... Sci[entific] or empirical interpretation from below upwards within the pyramid. Metaphysical explanation from above downwards and from an order of being outside the pyramid of nature. Creative and emergent evolution. (1921c: 5)

The above summed up Lloyd Morgan's own system. The metaphysical explanation of creative evolution started with deity, at the apex of the triangle, and proceeded top-down to space-time. The scientific interpretation of emergent evolution dealt with the three levels of materiality, vitality, and mentality inside the triangle, and proceeded bottom-up. Creativity was the activity of Source or Deity in the metaphysical realm, while emergence was the characteristic of evolution in the realm of nature. The highest level was that of mind, and it was here that Lloyd Morgan's combination of neutral monist and emergent evolutionary concepts caused the greatest difficulties.

(c) Emergence and Mind

Lloyd Morgan's discussion of mind involved a combination of emergent evolution and its view that mind emerged from organic matter, with neutral monism and its very different claim that the mental and the physical were correlated aspects of an underlying substance. Lloyd Morgan had defended a neutral monist view on the mind-body relation in his writings in the period from 1885 to 1912. What is significant is that he continued to defend neutral monism even after he developed emergent evolution. Neutral monism could take a number of different forms: from panpsychism and the claim that all physical entities, even inanimate ones, had a psychic correlate, to biopsychism and the view that only organisms, but all organisms, had psychic aspects, to zoopsychism which restricted the psychic to animals only. In volume one of his Gifford lectures, *Emergent Evolution*, Lloyd Morgan argued for a panpsychist correlation of the psychical with all physical things. However, this contradicted the emergentist principle of continuous stages. Of what use would life be as an intermediate stage in the hierarchy of matter, life, and mind, if mind appeared already as the correlate of matter, without the concourse or conditioning of life? This would amount to skipping a stage of evolution and omitting a level of reality, thereby violating the thesis of continuity. Yet in *Emergent Evolution* Lloyd Morgan said: "[T]here are no physical systems, of integral status, that are not also psychical systems; and no psychical systems that are not also physical systems." (1923: 25)

It was in volume two of his Gifford Lectures, *Life, Mind and Spirit* (1926), that he modified his position. The correlation, now termed "concomitance," was limited to the physiological and the psychological. This represented a retreat from panpsychism to biopsychism which was consistent with the principle of evolutionary continuity, though still not fully consistent with emergent evolution. Lloyd Morgan introduced the terminological distinction between "bioses," denoting any physiological system, and "neuro-bioses," denoting only neural systems. Unrestricted concomitance recognized that mental events accompanied all bioses, while restricted concomitance limited them to the accompaniment of neuro-bioses only. Lloyd Morgan opted for unrestricted biopsychism rather than the more restricted thesis of zoopsychism because he found it impossible to determine a criterion to distinguish between bioses accompanied by or concomitant with mental events, and those not so accompanied: "I am therefore prepared (until evidence to the contrary is forthcoming) to accept as a working hypothesis unrestricted concomitance of all bioses in an ascending order of rank with mental events in a corresponding order of rank." (1926b: 10-11)

Unrestricted concomitance, or biopsychism, was less universal in scope than the original panpsychist psycho-physical correlation, though it still posed a problem internal to Lloyd Morgan's system. According to this thesis even infusoria and amoeba, as living things, must have some psychic or mental life. The problem was to reconcile this position with the claim of emergent evolution that new qualities, in particular the mental, appeared in the course of evolution. At the level of unicellular organisms the emergence was that of life, with its novel properties of metabolism, growth, and reproduction, and this Lloyd Morgan readily admitted. The next step should have been to recognize that the psychological and mental had emerged in some, but not in all forms of life. But in discussing unrestricted concomitance, Lloyd Morgan did not extend emergence to mind, and to the contrary stated:

> Let it then be understood clearly that the hypothesis of unrestricted concomitance does *not* imply that occurrences in either attribute emerge from occurrences in the other. The hypothesis is that from the very beginning, so far as we can descry it, mind is concomitant with life. (1926b: 12, italics in the original)

To claim that the mental was concomitant with life was to diminish the significance of its emergence, and could be taken as the claim that mind was a mere epiphenomenon of life. Yet Lloyd Morgan persisted, though surely he must have seen the problem himself. And indeed he admitted not only a double meaning to 'mind' – the first two being mind as concomitant, and mind as emergent – but a triple one, as he indicated in *Emergent Evolution*: "We have seen that the word "mind" may be used in three senses: first, as Mind or Spirit in reference to some Activity, for us God; secondly as a quality emergent at a high level of evolutionary advance; and thirdly, as a psychical attribute that pervades all natural events in universal correlation." (1923: 37) Though the last phrase of this quotation should be modified, in line with the correction contained in *Life, Mind and Spirit*, to read "and thirdly, as a psychical attribute that pervades all physiological events in universal concomitance," there would still be three distinct concepts of mind in his system.[1]

It is possible to reconstruct Lloyd Morgan's motivation for this profusion of concepts of mind only by considering his general worldview. Science, via emergent evolution, postulated mind as emergent from life; philosophy, in the form of dual aspect monism, argued that mind was correlated with life as concomitant aspects of an underlying substance; and religion, as an interpretation going beyond science and philosophy, postulated a transcendent mind or God who was the Source of all activity, including evolution itself. Lloyd Morgan wanted to harmonize science, philosophy and religion in his worldview, which was more comprehensive than that provided by any of these three knowledge domains taken separately. In trying to develop this worldview, he grafted together three mutually inconsistent, though singly defensible, views. In other words, his worldview blinded him to his own inconsistency. A more cogent presentation of emergent evolution would dispense with the second and third meanings of mind, limiting itself to the concept of mind as emergent in some organisms, in particular, those which had evolved brains. The advantage of such a presentation would be economy of concepts and a greater consistency of structure, by recognizing emergent evolution as a philosophy of science, replacing and not supplementing neutral monism, and by excluding the religious concept of mind as deity as unnecessary for a science-based philosophy.

(d) Emergence and the Social

A missing component of emergent evolution as set out in the Gifford Lectures was that of society and social evolution. It was only in a separate pamphlet, *Eugenics and Environment* (1919), that Lloyd Morgan considered these problems from the emergentist point of view. Lloyd Morgan was preoccupied by typical eugenicist concerns: the fear that social evolution would result in the deterioration of the level of civilization, and the consideration of means to avoid this eventuality.

While Lloyd Morgan believed that there was a process of "deterioration of the race" – the decline in the average intellectual level of the population – he felt that social evolution might yet prevent the more serious "degeneration of the stock" – the lowering of the intellectual potential of the species. The basis for this idea was the distinction between biological and social evolution made earlier in his work. In biological evolution the agency of change was natural selection, which acted upon organisms to preserve the fit and destroy the unfit. But in social evolution, natural selection was no longer dominant. The fittest, the most intellectually capable, raised the level of the less fit, or intellectually less capable, by improving the level of the cultural milieu from which all, regardless of individual contribution, might benefit. Lloyd Morgan's solution to the problem of eugenics was to argue that while the eugenicist was right in seeing an actual deterioration of the race, there would be no degeneration of the stock so long as social evolution ran its progressive course.

But would not the intellectual and moral characteristics of the fittest be swamped by those of the unfit? In replying to this objection, Lloyd Morgan combined the concept of emergence with that of Mendelian inheritance to argue that new properties need not be swamped by the preponderance of older ones in the general population. Mendelian inheritance was particulate, with a characteristic either present or absent in progeny, and not "watered down in an intermediate blend." (1919: 61) As a result, a new character which emerged would not be lost entirely or be submerged by cross-breeding with those in whom it was lacking: "Apart from such occurrence of emergent characters, it is difficult to account for evolutionary progress, and the origin of what seem to be new departures." (1919: 61) So long as social evolution continued to select for the fittest ideas in the cultural milieu, these characters would be retained from generation to generation in at least some members of the population, providing an island of merit from which all could ultimately benefit. This essay of Lloyd Morgan's was unique in both discussing social evolution in emergentist terms, and in linking biological emergence with genetic theory.

(e) Evaluation and Problems

In his theory of emergent evolution, Lloyd Morgan provided a solution to the problem of qualitative novelty raised by Wallace, without, however, violating the canons of continuity and naturalism of Darwin. Lloyd Morgan combined an adhesion to naturalism concerning the world of matter, life, and mind, with the recognition that life and mind were irreducible to the chemical and physical, thereby avoiding both the dualism of the vitalists and the reductionism of the mechanists. This was an outstanding achievement, and the reason for subsequent interest in his system, which served as the seed or kernel for the development of a philosophical trend.

However, there were also problems in Lloyd Morgan's presentation of the theory which would provoke justified criticisms in the period to come. The major defect of his system was the inconsistency of combining neutral monism, in the modified form of unrestricted concomitance of life and mind, with emergent evolution, which argued for the emergence of mind from life in the course of evolutionary advance. This contradiction between Lloyd Morgan's ontology of neutral monism and his ontogeny of emergent evolution resulted in the further duplication of metaphysical evolution, a preformationist process where teleology was present and novelty absent, and physical evolution, an epigenetic process where teleology was not involved, and genuine novelty was present.

A second problem involved the theses of natural piety and *nisus* towards deity. Lloyd Morgan explained qualitative novelty as the outcome of a process of emergent evolution; the next step was to explain emergence itself. But Lloyd Morgan had no explanation within science itself, nor for that matter, within what he conceived of as his philosophy, since he accepted emergence with "natural piety." Ultimately the answer was theological, to be found in deity as the source of evolution. This was unsatisfying except to those who shared Lloyd Morgan's worldview which aimed at the reconciliation of science, religion, and philosophy. But on the more reasonable and economical assumption that emergent evolution should be a philosophy based solely on science, Lloyd Morgan's solution is put into doubt, as it will be by a number of succeeding emergentists.

A third problem concerned the status of the social. The topic was only sketched in the pamphlet *Eugenics and Environment*, and was absent in Lloyd Morgan's major work *Emergent Evolution*. Yet societies as systems of individuals linked by economic, social, and political relations, possess properties different from those of the biological individuals who compose them. The identification of the social as a level of reality will be the major contribution of the American emergentists R. W. Sellars and W. M. Wheeler. Lloyd Morgan established the basis for a new philosophical trend, and the following chapters will trace its subsequent development.

PART THREE

THE DEBATE OVER EMERGENT EVOLUTION

CHAPTER TEN

Other Systems of Emergent Evolution

Though Lloyd Morgan was the central figure in the development of emergent evolution, other thinkers turned to emergentism, and it became part of the philosophical patrimony – occasionally adopted, though in modified form, frequently commented upon, and sometimes rejected. Just as a number of traditions had converged on Lloyd Morgan, so the interpretation of his work diverged into a multiplicity of applications and developments. Emergent evolution became a philosophical trend, in the sense that it had a continuous series of defenders, commentators and critics.

Three major components of the trend of emergent evolution can be identified. Firstly, there was a core of thinkers who developed emergentist systems of their own. Like Lloyd Morgan, they placed emergence at the center of their philosophies, though they disagreed with him on other aspects of ontology, particularly as concerns the role of teleology and theology in evolution, and the identification of the levels of reality. Secondly, there was a peripheral group of sympathizers who incorporated the concept of emergence into their systems, though as a subordinate and not a central concept. Thirdly, there was a group of critics who considered emergentism to be sufficiently important to comment on it, though they ultimately disagreed with its viability or usefulness.

The "classical" period of emergent evolution was during the mid-1920s, as a number of other systems of emergent evolution were developed in book-length format, and the philosophy was widely debated. During the period from the mid-1930s to the early 1950s, emergent evolution was "eclipsed" as a philosophical trend by reductionism and the physicalist thesis that all properties could be reduced to the base level. Nonetheless, an aspect of emergentism was preserved in the concept of integrative levels in biology, and beginning in the mid-1950s, there was a general renewal of interest in emergent evolution and its claims for qualitative novelty and levels of reality. Emergent evolution has once more became an active area of research at the intersection of philosophy and science.

This chapter considers the other "core" theorists who developed versions of emergent evolution during the 1920s – Samuel Alexander, C. D. Broad, William Morton Wheeler, and Roy Wood Sellars. Debate among these theorists centered on the nature of the lowest, intermediary, and highest levels of reality, and the place of theological and teleological considerations in the evolutionary process. This diversity of views was not a sign of confusion but rather an indication of the vitality of emergent evolution as a philosophical trend.

(a) Pyramid of Levels: Samuel Alexander

Samuel Alexander (1859-1938) was a philosopher, originally from South Africa, who taught at Manchester, in England. His major work, *Space, Time and Deity* (1920) was presented as Gifford Lectures in 1915-16, six years before Lloyd Morgan's, but two years after this latter had published his first articles on emergent evolution. Alexander was not the only other lecturer in the Gifford series to advocate qualitative novelty in evolution, and was preceded in this by J. Arthur Thomson, who also was influenced by Lloyd Morgan.[1] Alexander readily acknowledged that his use of emergence was based on that of Lloyd Morgan: "I use the word "emergent" after the example of Mr. Lloyd Morgan. It serves to mark the novelty which mind possesses, while mind still remains equivalent to a certain neural constellation." (1920, 2:14) In turn, Lloyd Morgan was also influenced by Alexander, and adopted his term of "*nisus* towards deity." The interaction between Lloyd Morgan and Alexander had been a continuous one since the late 1880s, and the greatest number of letters from any correspondent in the Lloyd Morgan archives were those of Alexander.

Alexander argued that Space-Time was the basic level of reality, the stuff of which all things were constituted, the matrix or basis from which all the other levels emerged.[2] The concept of combined space-time was indebted to Einstein's theory of relativity and Minkowski's four dimensional continuum. But relativity provided at most the impetus for including the level of Space-Time as the primary one, and unlike Whitehead's work of the same period, Alexander did not explicitly take into account specific aspects of relativity theory. In Alexander's theory, Time was the determining force, the principle of motion and change, while Space was the passive partner, that which was acted upon. Taking the image of active and passive a step further, Alexander asserted that Time was the mind of Space, and Space was the body of Time. This mind-body metaphor was extended to inter-level relations, so that after the base level of Space-Time, the higher in each successive pair of levels played the role of "mind" to the lower, which had the status of "body" in relation to it. Alexander's use of the mind-body model to describe inter-level relationships can at best be considered as metaphorical, at worst as idiosyncratic. Though his stated reason for introducing it was to demarcate his system from materialism by incorporating mind throughout the system, it had the negative effect of multiplying the meanings of the term "mind" beyond that which psychology as a science recognized.

Alexander believed that Space-Time spontaneously took on motions, and that these motions were the basis for the emergence of things: "Empirical things come into existence, because Space-Time of its own nature breaks up into finites, the lowest such finites being simple motions of different velocities or intensities of motion." (1920, 2:47) From the primitive motions in Space-Time, under the lead of Time, developed the next two levels: the Primary Qualities and the Secondary Qualities. By placing the Secondary Qualities immediately after the Primary, and before the appearance of the level of human minds, Alexander rejected the traditional philosophical thesis that secondary qualities were mind-dependent. Yet Alexander's Secondary Qualities, though independent of human minds, had a "mental" role to play, as "mind" to the preceding level of Primary Qualities.

Alexander considered the possibility that the next level might be that of "chemism," the chemical organization of matter. This would have been in keeping with the original motivation for emergence, the belief that molecular organization featured new qualities in relation to the atomic, as evidenced in the recurring water, hydrogen, and oxygen example. But Alexander believed that the distinction between Primary and Secondary Qualities, rather than that between molecules and atoms, better fit the mind-body model of the relationship between higher and lower levels. As a result, he omitted the chemical level and postulated Life as the next level, immediately followed by that of Mind.

With the level of Mind appeared the tertiary qualities of truth, goodness, and beauty. Unlike Primary and Secondary Qualities, tertiary qualities did not constitute a distinct non-mental level. Rather, they depended on minds and their interaction with non-mental objects: "Strictly speaking it is the totality of knower and known, of subject and object, which is true or good or beautiful." (1920, 2: 239) Tertiary qualities were therefore "subject-object determinations," differing only in that truth was determined primarily by the object, while beauty and goodness were determined primarily by the subject.

The highest level was that of Deity, a level which was infinite yet still in the process of formation. Deity served as the mind of all the previous levels, introducing the third sense of the term "mind" in Alexander's system. Alexander's Deity, unlike Lloyd Morgan's Source, was not apart from the levels of reality. Reality tended to Deity as its supreme evolutionary product, a concept captured in Alexander's phrase "*nisus* towards deity': "There is a *nisus* in Space-Time which, as it has borne its creatures forward through matter and life to mind, will bear them forward to some higher level of existence. ... Deity is the next higher empirical quality to mind, which the universe is engaged in bringing to birth." (1920: 2: 346-47) The following is a graphical representation of Alexander's system, applying Lloyd Morgan's technique of representing the hierarchy of levels as a pyramid. Italicized terms are used for levels, or parts of levels in the case of Time, which serve as "minds" for other levels, or parts of levels in the case of Space:

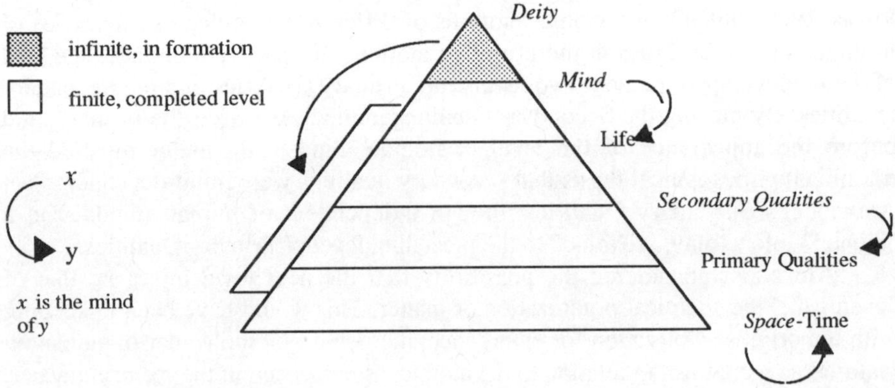

Diagram 10.1: Graphic Representation of Alexander's Level Structure of Reality

What remained constant throughout this maelstrom of level-building activity were the categories, beginning with the first two "ranks" or levels of Identity, Diversity, and Existence, and Universal, Particular, and Individual, and building up through a series of levels of further categories to that of Motion, which was the key to the dynamic elaboration of the levels of reality described previously. Typical of Alexander's reasoning was his discussion of the first two groups of categories. A "piece" of Space-Time was Identical to itself and different or Diverse from other pieces. It manifested the category of Existence or determinate being as the result of the union of its Identity and Diversity. Universality was the identity of kind, and Particulars which shared a kind fell under the same Universal. Universal and Particular, however, were one-sided, and found their real existence in pieces of Space-Time united as Individuals. This process then was continued through extended discussions of the other categories.[3]

However, the transitions were often abrupt and contrived, particularly between Number at the seventh rank of categories, and Motion, the eighth and highest rank, where the shift from the mathematical to the dynamic occurred. Though the categories were to be the "fundamental determinations" of Space-Time, they suffered from Alexander's *a priori* attitude – his refusal to take into account the results of the science of space-time, as formulated by relativity theory. This is not to say that philosophy should be merely a deduction from science, but the two are related, particularly for a system of emergent evolution which took a discovery of natural science – evolution – as the basis for a metaphysical system. Alexander disconnected science and philosophy in his treatment of space-time, and as a result his theory of categories was artificial and unconvincing.

Alexander's ontology was monistic. In his structure of levels, a higher level was, insofar as stuff was concerned, "expressible completely without residue" in the lower, and ultimately in terms of Space-Time events. There was nothing substantial added to get from one level to another. What changed from level to level were the qualities which emerged, and which were associated with higher forms of organization of spatio-temporal stuff.

Alexander opposed his monism to materialism, which he termed a "word of abuse," and claimed to avoid in two ways: in the first place, matter itself was not primitive, but the outcome of intrinsic motions of Space-Time, and secondly, mind was present at the earliest level, at least implicitly, as Time. Time *qua* mind initiated those motions which led to matter: "My motive in anticipating the discussion of empirical qualities by the hypothesis that Time performed towards Space the office of mind, was, that by suggesting that something corresponding to mind was present from the beginning at the lowest level of mere motion, I might remove the prejudice against any attempts to exhibit all forms of existence as a continuous series from Space-Time upwards through matter to mind." (1920, 2: 50). This critique of materialism was based on the belief that only eliminative materialism was possible, and that such a philosophy, because it rejected mind, was unacceptable. However, materialism, like idealism and dualism, can take on many forms, including an emergentist variant recognizing the existence of mind, which philosophers such as Roy Wood Sellars and Mario Bunge would defend.

(b) Mechanism, Vitalism, and Teleology: C. D. Broad

The defence of emergence, albeit temporary, by the English philosopher C. D. Broad (1887-1971) marked an inroad of that philosophy within a different intellectual trend – that of analytic philosophy – of which Broad, along with Bertrand Russell and G. E. Moore, was a foremost representative. Broad considered emergentism as the alternative to the extreme positions of mechanism and vitalism, avoiding both the reductionism of the former and the dualism of the latter. Mechanism was too simplistic a theory since it claimed that all things were made up of parts of a similar nature, affected only by changes in position and pair-wise interactions through contact. Broad argued that mechanism broke down when applied to secondary qualities, since the immediate consciousness of sense-data could not be reduced to the arrangements of moving particles in external objects.

Broad reinforced his rejection of mechanism by further arguing that the concept involved a hidden teleology, where teleology was taken to mean intelligent design for a specific purpose. If biological mechanism as a philosophy were true, then all living bodies would be machines. But a machine was designed for a purpose by entities of a higher order than it – humans in the case of artefacts. Now, if humans were machines, there must also be a designer of a higher order than humans, a super-human designer: "Thus the proper complement to a completely mechanistic theory about organisms is some form of the doctrine of Deism." (1926: 90). Broad, however, wanted his natural philosophy to be independent of any theological commitment, and this became a further reason motivating his rejection of mechanism.

Broad excluded the other extreme of vitalism as overly complicated and inconsistent with scientific method, that of chemistry in particular. Vitalists called for a special, non-material component to explain the existence of living organisms, and Broad took as typical Hans Driesch's "entelechy," an immaterial source of life similar to Bergson's *élan vital*, but specific to each body and inseparable from it. Driesch defended this concept in his Gifford lectures of 1907-08, *The Science and Philosophy of the Organism*. But Broad argued that unlike chemical radicals, entelechies could not be isolated from the body they inhabited, or be transferred from one body to another; nor could anything be said concerning the manner in which the entelechy differed from the body, or the overall structure of the organism-entelechy interaction. This ruled out vitalism as an adequate ontology for a scientific philosophy, since it postulated a basic entity the existence of which could not be experimentally confirmed or disconfirmed.

Having excluded the mechanistic and vitalistic extremes, Broad adopted the emergentist option, stressing three characteristics: firstly, the biological whole was determined by the nature and arrangement of its component parts, and no special, non-material component or life principle was required; secondly, the behaviour and properties of the biological whole could not be predicted from that of the behaviour and properties of the component parts taken separately; and thirdly, the law relating the whole and its parts was unique and irreducible, not a special case of a more general law or a combination of two or more general laws. Broad also distinguished between three types of properties: emergent properties which were ultimate properties characteristic of an order, since they did not exist at lower levels, and could not be deduced from the properties of any lower order or orders; reducible properties which could be deduced from knowledge of the properties at lower levels; and neutral properties which existed at more than one level of the hierarchy. Broad's examples included sexual reproduction as an emergent characteristic of the biological order, since it did not occur in non-living things; the beating of the heart as a reducible property, since it could be derived from the strictly mechanical properties of the heart muscles; and the conservation of energy as a neutral property of both living and non-living matter.

Broad concluded that the physical, the chemical, and the biological were the first three orders or levels of reality, with each one emergent relative to the preceding. This left the problem of the status of the psychological relative to the physiological. In analyzing this question, Broad developed a classification schema for the mind-body problem which was very extensive, and which included a total of seventeen different theories as candidates. In his schema, Broad distinguished three types of attributes: differentiating attributes which determined distinct substances, as opposed to non-differentiating attributes which merely distinguished kinds within the same substance; applicable attributes within the non-differentiating ones, as opposed to those which were not applicable to anything, termed "delusive"; and emergent attributes among the applicable non-differentiating ones, as opposed to those which were reducible as they could be derived from other attributes. The relationships among these types of attributes can be represented as follows:

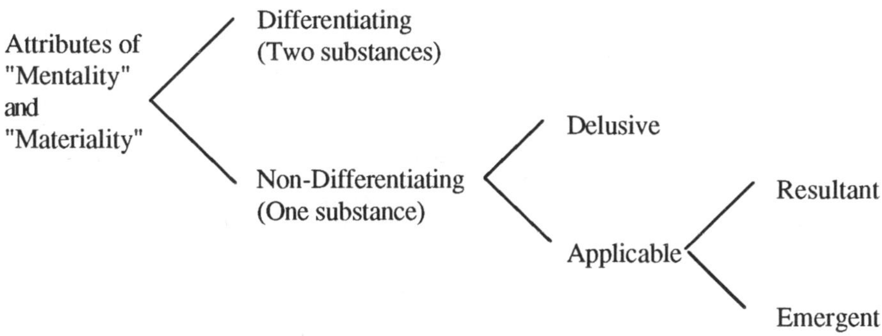

Diagram 10.2: Broad's Categories for the Classification of the Mind-body Problem

Applied systematically, these three groups of attributes resulted in a total of seventeen different theories of the mind-body relation. The main branching was the result of three possibilities for the differentiating or non-differentiating status of "matter" and "mind." If both were differentiating, dualism resulted. If only one were differentiating, the result was either materialism or idealism, and in the case whether neither were differentiating, neutralism resulted. The second level of analysis considered the applicable or non-applicable distinction. A pure form of materialism, idealism, or neutralism resulted when the non-differentiating term was delusive. Where the non-differentiating term was applicable, the emergent-resultant distinction was applied as a third level of analysis, resulting in two possibilities for each of materialism and idealism, and eight special cases for neutralism. A chart of the classification proposed by Broad can be constructed as follows:

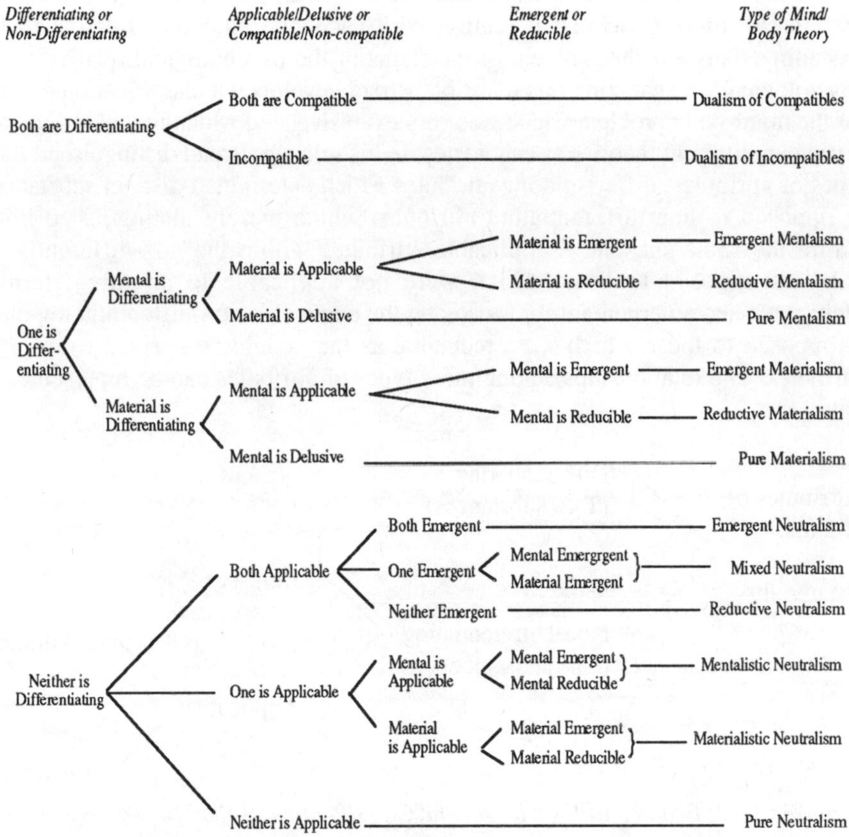

Diagram 10.3: Schema of Broad'sAnalysis of Mind-Body Philosophies

Just as dualism was under-represented with only two types recognized – the dualism of compatibles and the dualism of incompatibles – so was neutralism over-represented with a plethora of types, many of which had never been defended in the history of philosophy. The analysis of materialism and idealism was of greater import, and set the stage for Broad's choice of an option to defend.

Broad's conclusion that the psychological order was also emergent was consistent with the way he had structured the previous levels, given that the physical, chemical, and biological were analyzed in emergent terms. As a result, he defended an emergent materialist solution to the mind-body problem, holding that "(a) materiality is a differentiating attribute, and (b) that mentality is an emergent characteristic." (1926: 646) Broad was the first of the emergentist theorists to adopt such a materialist ontology, but he added the proviso that this held only so long as consideration was limited to what he termed "normal" facts: "If there were no facts to be considered except the normal ones, and we rejected all the alleged abnormal facts dealt with by Psychical Research, I should regard Emergent Materialism as on the whole the most reasonable view to take of the status and relations of matter and mind in Nature." (1926: 648)

But despite his desideratum of keeping philosophy distinct from theology and its commitment to the supernatural, Broad took seriously the "alleged abnormal facts" of parapsychology, and recognized the existence of a different kind of supernatural entity, "something which is capable of persisting for some time after the death of the body and of entering into temporary combination with the brain and nervous system of certain peculiarly constituted human beings called 'mediums'." (1926: 651) Broad stated that there was evidence for the survival of the psychic factor for a few days following the death of the body, though he did not elaborate on the details. He avowed that were there evidence for permanent survival, he would be a dualist, but given his assumption of only temporary survival, he shifted to an intermediate or "Compound Theory" point of view. Mentality was defined as "an emergent characteristic of a compound of a living brain and the psychic factor." (1926: 651) In his subsequent writings on the nature of mind, Broad dealt exclusively with the paranormal psychic factor and dropped further reference to emergent materialism.[4]

This development was inconsistent with Broad's own criticisms of entelechy and the vital principle in biology, where he had convincingly argued against the addition of any such non-material factor. Broad had excluded the vital factor because it could not be isolated, transferred, or analysed. The same applied, however, to the psychic factor. Broad did not state how it might be transferred, isolated, or analyzed. Why did it persist for only several days and not several weeks? What did it decompose into after its period of persistence was over? How was it related to the mind in a living person? How did it combine with the body of the medium? These questions were and remain unanswerable, and because of that, Broad's compound theory was inadequate. The strength of his defence of emergentism, however, was in situating it as an alternative to both mechanism and vitalism.

(c) The Emergence of the Social: W. M. Wheeler

The Harvard entomologist W. M. Wheeler (1865-1937) was one of the discussants of emergent evolution at a colloquium held on that subject at the 1926 World Congress of Philosophy. Wheeler's article "Emergent Evolution of the

Social" was reproduced twice, once in *Science*, and then as a pamphlet entitled *Emergent Evolution and the Development of Societies* (1928), with an additional chapter relating emergentist and organicist ideas. Wheeler's interest in the social insects was the immediate background to his discussion of societies as a level of reality. He noted that comparative sociology was neglected since the sociologists left insect and animal societies to the biologist, while the biologists were little concerned with the collective dimensions of life, limiting themselves to the characteristics of typical individuals. But nonhuman societies "no less than human society, are as superorganisms obviously true emergents, in which whole organisms function as the interacting and determining parts." (1928: 36) Societies possess properties not possessed by any of the organisms which compose them, constituting an emergent level of reality after those of the material and the biological.

In his taxonomy of social aggregates, employing the term "social" in the widest sense, Wheeler distinguished between homogeneous and heterogeneous societies. In homogeneous societies reproductive or defensive reasons predominated in the emergence of social behavior and the division of labour among the population. These groups "constitute a vast series of emergents varying from those of very low to those of very high integration." (1928: 37) Heterogeneous social aggregates were systems of interacting species based on relations of predation and parasitism, as well as symbiosis and association. Wheeler applied the concept of "heterogeneous emergence" to multi-species ant colonies, where one of the species "enslaved" the other. In this case, there was role differentiation not only within the species, but between species, "so that a new emergent arises – a super-superorganism, or superorganism of the second degree." (1928: 40) The concept of heterogeneous emergence could also be extended to cases of symbiotic associations between different species, where long term arrangements of mutual benefit brought about the emergence of new properties in the host-symbiont pair.

Because of the pervasiveness of the social, Wheeler concluded that it was not only an emergent from life, but was correlated with it as well: "We may say, therefore, that the social is a correlate as well as an emergent of all life in the sense in which [Lloyd] Morgan speaks of the mind as being both a correlate and an emergent of life." (1928: 41) Wheeler extended the correlation of the social even lower to the inanimate, believing that the structure of the atom involved the association of electrons and other constituent particles in a type of "social" organization. In over-extending the domain of application of the social Wheeler adopted the "socialist" equivalent of an organicist position: the social was a basic characteristic not just of one level of reality, but of all. The same criticism that was applied to Lloyd Morgan's correlation of the mental and the organic applies to

Wheeler's correlation of the organic and the social: it diminished the importance of the emergent and tended to equate it with the level from which it emerged, at least in terms of extent.

Wheeler held that emergence was a strictly natural process, requiring no immaterial entity outside of time and space, in explicit opposition to Driesch's entelechies and Bergson's *élan vital*. He also rejected any reference to a supernatural entity: "I fail to understand why Alexander and Morgan select deity as the supervenient level next to mind, since their general scheme of emergent evolution most naturally demands the social as the next level in ascending order." (1928: 41-42) Wheeler noted that the next emergent level after that of the social could not be predicted, though he entertained the possibility that there might be an end to the series, or "supervenient extinction."

A danger which disturbed Wheeler was that of the degeneration of the individual as the result of extreme social organization, a tendency he believed to exist in the social insects. He was concerned that this might occur in humans too, and mentioned signs such as the decline in the acuity of the sense-organs, absence of demonstrable increase in average intelligence during historic time, and the increase in mental illnesses and social disorders in large cities. He wondered whether this might not lead, as had been the case with the social insects, to "a society of lower intelligence of the individuals combined with an intense and pugnacious solidarity of the whole." (1928: 43) Further social development of this sort would result in biological inadaptation and eventual extinction, and Wheeler concluded on this pessimistic note.

Wheeler was the first supporter of emergentism to include the social as a level, and in his discussion of the relation between the biological and the social, he developed many of the themes later taken up by E. O. Wilson in his sociobiology. Indeed, Wilson mentioned Wheeler as the precursor of his thought in both his *Insect Societies* (1971) and his *Sociobiology: The New Synthesis* (1975). However, whereas Wheeler recognized that higher levels could not be reduced to lower levels, a tendency in sociobiology has been to reduce the social to the biological by minimizing the differences between human and non-human societies.[5]

(d) Evolutionary Naturalism and Materialism: R. W. Sellars

The early work of the philosopher Roy Wood Sellars (1880-1971) stressed naturalism in ontology, critical realism in epistemology, and humanism in ethics, themes which were to be a constant throughout his life. In *Evolutionary Naturalism* (1922), he stressed the importance of novelty in the relation between the inorganic and the organic: "The problem between science and philosophy in these days is to harmonize new forms of organization, new types of behavior, new properties with the more stable and fixed background of inorganic nature." (1922: 263) But the terms for his discussion of novelty were the traditional Aristotelian ones of potentiality and actuality, and necessity and contingency, rather than the emergentist ones of part-whole and qualitative novelty. In his exposition of an evolutionary philosophy, Sellars focused on the naturalistic solution of the mind-body problem, and considered that the key thesis of his naturalism was that the living organism included consciousness as an integral aspect, and was its sole source. As of yet, there was no clearly formulated concept of levels of reality, an essential ingredient for any emergentist philosophy.

It was in *Principles and Problems of Philosophy* (1926) that Sellars made explicit his adoption of an emergentist point of view. Referring to the hypothesis that there was novelty in the world, he noted: "This thesis has been given various names of much the same import: creative evolution, emergent evolution, epigenetic evolution, originative evolution." (1926: 362). Sellars agreed that higher level laws could not be deduced from lower level ones, so that chemical laws, for example, were not deducible from mechanical ones, and biological laws were not deducible from chemical ones. Such laws must be discovered, not inferred, and "form a hierarchy in which the different levels are discontinuous." (1926: 364) But at the same time, Sellars stated that evolution itself was not discontinuous and did not require saltationist jumps. Rather, there were "mutative junctures" when novel properties appeared which led to new levels of reality. These mutative junctures must be accepted with "natural piety," a proposal which did not contribute to solving the problem of combining discontinuous levels and continuous evolution. As in the case of Lloyd Morgan and Alexander who originated the term, natural piety put a premature stop to further philosophical analysis.

Sellars believed that increase in complexity of organization was the driving force behind emergent evolution. Emergent properties were functions of organization and not due to metaphysical or supernatural influences. This was represented by the upward pointing arrow labeled "organization" in the diagram, and replaced the "*nisus* towards deity" of Alexander and Lloyd Morgan. Sellars also differed with Lloyd Morgan and Alexander since he argued that evolution was not teleological, and that emergence was only a local, not a universal phenomena. In "Reformed Materialism and Intrinsic Duration" (1944) he rejected both a linear view of evolution which pictured an initial creation followed by universal and directed development in time, and the opposite notion of the running-down of the universe and its eventual exhaustion. Instead, he opted for an eternal universe with no universal progress towards a fixed and pre-determined end, one where life and mind emerged only locally and but occasionally: "I am naturally led to suppose that the universe has always been much as it is now, a variegated existential domain with a floor, much the same everywhere, above which rise here and there mountain peaks of emergent becoming followed in time by recession." (1944c: 380) A similar point had been made earlier by A. O. Lovejoy in "The Meanings of 'Emergence' and its Modes" (1926), which had been presented at the debate on emergent evolution at the VIth World Congress of Philosophy. Sellars identified matter, life, mind, and society as the levels of reality. Like Lloyd Morgan, he represented the emergent levels of reality by a pyramid drawing: (1926: 345)

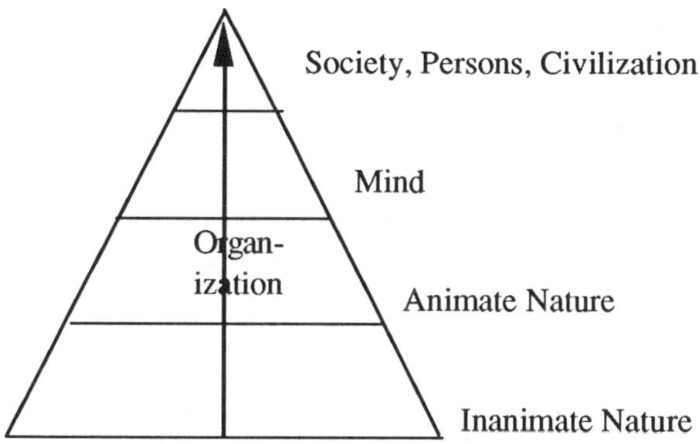

Diagram 10.4: Sellars' Representation of his Hierarchy of Levels

Sellars was opposed to materialism in its older forms, which he associated with a mechanistic approach to vital and mental phenomena, a pre-evolutionary conception of change as mere re-arrangement, an atomistic view of the nature of things, a naïve realist epistemology, and a lack of developed axiology. Yet his ontology began with matter and rejected all supernatural agents, and Sellars moved to explicitly adopt a new, or emergent materialism in a series of subsequent articles. In "Why Naturalism and not Materialism?" (1927), he distinguished between naturalism as a more general, cosmological view and materialism as a more restricted, ontological position, consistent with and subsumed under naturalism. But in "Is Naturalism Enough?" (1944), he answered the question in the negative and advocated emergent materialism as a general metaphysical position as well. This was extended in "Can a Reformed Materialism Do Justice to Values?" (1944) to a defense of an emergentist approach to axiological problems. Consistent with this change in metaphysics, Sellars was one of the co-editors of *Philosophy for the Future: The Quest for Modern Materialism* (1949). This was one of the few volumes in the United States of the late 1940s and early 1950s to explicitly advocate philosophical materialism, in defiance of the growing McCarthyism and its confounding of materialism and communism.

(e) Debate over the Levels of Reality

The question which was uppermost in the debate among the five original authors of emergentist systems – Lloyd Morgan, Alexander, Broad, Wheeler, and Sellars – was that of the identification of the levels of reality. This controversy can be divided into three parts as follows: What was the basic or first level? What were the intermediate levels? What was the highest level? A comparative table of the levels identified by each of the authors is the following:

level	*Lloyd Morgan*	*Alexander*	*Sellars*	*Wheeler*	*Broad*
space-time		1			
matter	1		1	1	
primary qualities		2			
secondary qualities		3			
physical					1
chemical					2
life	2	4	2	2	3
mind	3	5	3		4
society			4	3	
deity		6			

Table 10.1: Comparison of Levels of Reality

Alexander argued for space-time as the first level, while Sellars, Broad, and Wheeler opted for matter. Lloyd Morgan adopted a compromise view, with matter as the first level, but with space-time as the stuff from which material things were made. Alexander believed that, based on the categories, motions in space-time built up the levels of reality. Taken to an extreme his view implied that the space-time motions replaced evolution as the driving force of level building, a downgrading of the importance of evolution which did not occur in the other systems. Space-time is a form, while matter is a content, and Alexander was unable to convincingly argue how the former could produce the latter. Beginning with matter as the first level of reality has the advantage that higher level systems are readily built up of its units – the atoms – which combine to form molecules, macro-molecules, cells, and organisms, of which some develop minds.

Life and mind were included among the intermediary levels by all of the emergentists, with the exception of Wheeler, who was mainly concerned the relation between the biological and the social. The core position of emergentism was typically formulated by Lloyd Morgan, who considered that matter, life, and mind were the essential levels. However, he conceived of matter and life as physical with psychological concomitants, and mind as psychical with a physical concomitant. A more consistent presentation of the three levels would build up the two higher levels from the material one based on emergent evolution alone, without reference to neutral monism. The material level of inanimate, unthinking atoms and molecules would give rise to the further physiological and psycho-physiological levels. Life would be the level of cell-based organisms characterized by the emergent properties of nutrition, metabolism, and reproduction, while mind would be the level consisting of organisms in which brains had evolved, and which as a result possessed the emergent properties of perception, consciousness, and intelligence.

Differences were greatest concerning the highest level, where Alexander argued for deity, Lloyd Morgan and Broad opted for mind, while Wheeler and Sellars argued for society. Within a naturalistic framework, the question to be resolved was the relationship between mind and society. The further question of the status of deity arose only in systems that admitted a supernatural component in addition to a natural one. The disadvantage of such a system was, as Sellars and Wheeler argued, the coupling of theological with teleological considerations. Deity, as first and final cause, as source and end of the evolutionary process, was neither justified nor required by science and a science-based philosophy. Moreover, acceptance of a teleological component weakened the role of emergentism as an alternative to both mechanism and teleology, as initially pointed out by Broad.

The advantage of a system such as that of Sellars was that only one type of entity was needed – the objects of science – making the system more compact than those which included divine entities as well. However, the difficulty that Sellars faced concerned the relation between the mental and the social: on the one hand, though the social followed the mental, societies did not possess mental characteristics, and on the other hand, minds, though preceding the social, were socially conditioned in that individual minds were influenced by their surrounding social circumstances.

Placing the social before and not after the mental could resolve these two difficulties. Mind would jointly depend on the preceding biological and social levels, as originally suggested by Lewes. The emergence of the mental would presuppose not only the neuronal organization of the cerebral cortex which had evolved in some species of organisms, the animals in particular, but also the social organization of those organisms. The development of societies has resulted in division of labour and systems of communication which have spurred the development of mind. The mental would then be an emergent property of some socially organized and cerebrally endowed organisms, a thesis which will be further examined in the conclusion.

CHAPTER ELEVEN

Systems Related to Emergent Evolution

Further attempts to develop concepts of emergence were made by a number of authors in the 1920s and early 1930s. George Conger analyzed emergence as the "epitomization" of later structures by earlier ones, Oliver Reiser considered emergence to result from stochastic processes, W. P. Montague argued for emergence as the product of the interaction of two factors at each level, and George Herbert Mead saw emergence as the co-presence of a thing in different systems, conditioned by an ever present "sociality." Of interest as well are two other systems of the 1920s – J. C. Smuts' "holistic evolution" and J. E. Boodin's "cosmic evolution" – and two systems of the 1940s – the ontologies of Nicolai Hartmann and James Feibleman. Common to all these philosophies were attempts to analyze emergence in terms of more basic concepts, combined with an acceptance of the emergentist concept of irreducible levels of reality.[1]

(a) Emergence as Epitomization: George P. Conger

George P. Conger's concept of epitomization in "Evolution and Epitomization" (1921) was related to Spaulding's concept of "creative synthesis" in *The New Rationalism*, which had appeared two years earlier. Following Spaulding, Conger held that actual integrations were not mere aggregations, but systems which possessed novel qualities. A whole was equal to the sum of its parts only as an abstraction in mathematics, not in evolutionary reality. Conger initially defined epitomization as a process of development in which later stages recapitulated, with evolutionary novelty, earlier stages. Each stage or unit of development was termed a "monad," and evolution was analyzed as a series of monads. The system was monistic since no force other than the natural evolutionary process was required to move from one series to the next. Conger included three series: the physical series, which went from sub-atomic particles to galaxies, the biological series, including single and multi-celled organisms as well as societies, and the mental series, also called the neuropsychological, which proceeded from simple effector and receptor cells to sentiments, values, and selves.

Conger further elaborated his philosophy in "The Doctrine of Levels" (1925), bringing it into line with the emergentist terminology of Lloyd Morgan. The publication of this article was the occasion for an exchange of views with Harold Chapman Brown (Stanford University). In "A Materialist's View of the Concept of Levels" (1926), Brown, who was opposed to both mechanism and teleology, also rejected emergentism and the thesis of the existence of three separate series or levels of the physical, the biological, and the mental. Rather, he argued that the major bifurcation within the natural world was within the biological, between unicellular organisms, derived via evolution from non-living matter, and multicellular organisms, among which mind developed in the course of further evolution. Mind could be fully explained as the result of the action of biochemical complexes already present in the earliest multicellular organisms, so that no new property emerged when the mental appeared. Though Brown considered the possibility of a third bifurcation, that of the social, he deemed it more probable that social processes were just a statistical manifestation of the behaviour of the biological individuals belonging to extended groups.

Conger replied in "What are the Criteria of Levels?" (1926) where he reviewed his analysis of the three series of the cosmogonic, biological, and neuropsychological, with a total of nineteen levels each, for a total of fifty-seven. In the physical realm, the advance was from aether and space-time through to the physical cosmos; in the biological, from the inorganic environment of life through to the totality of nations forming the "great society;" and in the neuropsychological, from animal minds to the totality of values or ideals considered as personality. It was not surprising that Brown, who had difficulties with three levels, ended the exchange of articles at that point. Though the views of Conger and Brown were not antagonistic in principle, they failed to coalesce because of the extremes of complexity and simplicity each defended. Conger proposed an emergentist theory of levels of reality, but one so complicated and eccentric as to discourage adherents, while Brown advocated a materialist point of view, but one too simplistic when it came to the levels of reality, which he reduced to the simple bifurcation between uni-cellular and multi-cellular organisms.

Conger, unconvinced of the need to simplify his system, continued to multiply the number of levels and series, which he extended in "A Hypothesis of Realms" (1928) to the formal sciences of mathematics and logic. The physical, the biological, and the mental series were now called "realms', but the substantive modification was the addition of three further realms from the formal sciences: logic, mathematics, and "geometry-kinematics." Adopting a Platonistic view, Conger considered that mathematical and logical entities were "independent not merely of mind, but of life and matter as well." (1928: 209) The third abstract realm, that of the entities or events of geometry-kinematics, was added to a provide a transition from the logico-mathematical to the factual realms: "My conclusion, then, is that there is a chance for a consistent argument that logic, number, geometry-kinematics, matter, life, and mind form a progressive series of realms offering at least a framework for a realistic metaphysics to be developed in accordance with the data of the various sciences both natural and abstract." (1928: 216)

This plan was carried out in *A World of Epitomizations: A Study in the Philosophy of the Sciences* (1931), a massive work of over 600 pages with sections numbered down to four decimal places; a summary was contained in the article "Epitomization and Epistemology" (1933). Conger noted that some structures and processes from one realm persisted into later realms and that this allowed for elements of reductive treatment: "This is why psychology, for example, taken analytically or reductively, is after a fashion describable in terms of physiology, and physiology in terms of physics and chemistry, and physics and chemistry in terms of mathematics and logic." (1933: 74) The term "epitomization', once used to encompass both recapitulation and novelty, was now reserved for the former alone, and the term "cumulative coordination" was introduced to refer to the novelty of later stages as compared with earlier ones: "Cumulative coordination makes for differences between successive levels and realms, while epitomization makes for underlying and essential resemblances." (1933: 74) This was similar to Lloyd Morgan's theory of resultant properties supplying continuity for qualitative novelties in the evolutionary process.

But the damage done by Conger's multiplication of levels was such that his system, despite a second edition of the book in 1944, was of no influence on other thinkers. Nonetheless, Conger could be quite clear and expository. His *New Views of Evolution* (1929) was a useful history of evolutionary philosophy of both the 19th and 20th centuries, and one of the few books by an emergentist that criticized the campaign against evolutionary theory being waged in the southern United States during the period of the 1920s.

(b) Emergence and the Stochastic: Oliver L. Reiser

For Oliver L. Reiser (1895-1974) emergence rested on the transition from the macroscopic to the microscopic. This concept was broached in his first article on the subject, "Creative Monism" (1924), and was further developed in "An Electromagnetic Theory of Matter, Life, and Mind" (1925) where he summed up his thesis as follows: "The theory which I wish to propose is that these discontinuities of the strata of nature, with their new laws of behavior, emerge when the microscopic whirls of events are unified into the groups of behavior which may be termed macroscopic rhythms." (1925: 611-612) In the transition from the micro-world to the macro-world, statistical considerations were brought into play, *a priori* predictions could not be made as to the exact outcome, and empirical observations became the chief means of making discoveries. In

"Probability, Natural Law, and Emergence" (1926) Reiser distinguished between causal and statistical laws of nature: causal laws gave "rigid determinism and predictability," while statistical laws "yield mere probability and introduce indeterminism into the calculations." (1926: 428) The transition from the microscopic to the macroscopic brought about "order" from "chaos" through a process of statistical averaging, resulting in the emergence of novelty: "Now if order can be produced from the "chaos" of random distributions through an averaging process, perhaps the "emergents" which are figuring so prominently in the speculations of contemporary evolutionary philosophy are but the "central tendencies" of statistical distributions." (1926: 431) The notion of the transition form chaos to order has recently been more thoroughly studied in the theory of non-linear dynamics developed, among others, by Ilya Prigogine, and has given rise to a new area of study, that of "chaos" theory.

In his "Mathematics and Emergent Evolution" (1930), Reiser stated that the difference between emergent and resultant processes required the use of two different types of mathematics. Continuous equations of differential calculus applied when working within a level of reality, but discontinuous, finite difference equations must be used in going from one level to another. In *Philosophy and the Concepts of Modern Science* (1935) the distinction was expressed as follows: additive properties "Wadd." were represented by a linear equation: "*Wadd.=f(A) + f(B) + f(C) + ... +f(N)*," while emergent (non-additive or gestalt) properties "Wnon-add." were non-arithmetical functions of their parts: "*Wnon-add.= f(A, B, C, ... ,N)*." (1935: 131) However, Reiser became interested in organicism as a philosophy of biology, and evolutionary humanism in ethics, and devoted his further research to these questions rather than to emergentism.

(c) Factors of Emergence: W. P. Montague

William Pepperill Montague (1873-1953) in "A Materialistic Theory of Emergent Evolution" (1929) attempted to provide a mechanism to explain the emergence of plants, animals, and persons. He argued that this could be done by demonstrating the existence at each level of reality of two factors, one dominant and the other subordinate, such that the emergence of novelty occurred when the subordinate factor of a previous level became the dominant one of a new level: "That is to say, we shall endeavor to show that there are the same two factors in any pair of successive levels, and that the transition from one to the other, momentous though it may be, is adequately explained by a continuous quantitative growth of the lower factor until the critical point is reached when it gains ascendancy over its former superior and there emerges a novel quality of being seemingly discontinuous with that which gave it birth." (1929: 269)

At the inorganic level the dominating factor was that of kinetic energy "with its tendency to diffuse and run down by conduction," while the secondary factor was that of "potential energy with its tendency to perpetuate and reproduce by polaric induction." (1929: 269) With the evolution of protoplasm, this second factor came to predominate and with it emerged the organic level. At the level of plant life the dominant factor was "the capacity for receiving, retaining, and organizing the energies that are bound up with matter as food" (1929: 270), while the secondary factor was the same capacity, but extended to the active utilization by the nervous system of energies, such as light and sound, for information processing and subsequent action. Animal life emerged with the dominance of the second factor, so that the animal was a "doer," and not merely a "grower" as with the plant.

At the animal level, the factors identified were "the vegetal and inherited factor of body-building anabolism," and "the private history of the organism's own adventures with the surrounding world." (1929: 271) The dominance of this second factor, which included the consciousness of the organism of its own life, led to the personal level. Montague also admitted the possibility of a fourth emergence, from the personal to the divine, with the divine considered as human culture, expressed metaphorically as follows: "If man were able to give actual life to the children of his spirit and to endow them with the same independence and capacity for self-preservation and reproduction as that possessed by the parts of his body, he would have attained the next higher level of evolution, which following Alexander, we may denominate the level of Deity." (1929: 272)

Here, as in his description of the other factors, Montague's description tended to be more picturesque than exact. A further problem concerned the quantitative determination of the critical point which once achieved led on to emergence of novelty. How could the various factors be quantified? How could the value of the critical point be determined? Montague, like Sellars before him, made no suggestion concerning a measure of complexity that might be relevant to this problem.

(d) Emergence and Sociality: G. H. Mead

George Herbert Mead (1863-1931) in *The Philosophy of the Present* (1932) was concerned with the problem of time, and in particular, the relation of past, present, and future. As suggested by the title of his book, he privileged the present. The matter was stated quite boldly at the outset: "The subject of this lecture is found in the proposition that reality exists in a present. The present of course implies a past and a future, and to these both we deny existence." (1932: 1) This belief in reality as only a present was combined with the acceptance of the emergence of novelty. Emergent events were what determined the perception of the present, so that the chief reference of a present was to an emergent event, and at the instant of its appearance there was a break with the past. But in Mead's view the occurrence of a novel event brought about a rewriting of the past so that this breach with continuity was done away with: "The emergent then ceases to be emergent and follows from the past which has replaced the former past." (1932:11)

It was a startling proposition that the present "rewrote" the past, and in so doing, the present emergent ceased to be emergent – shades of George Orwell's *Nineteen Eighty Four*. But Mead argued that this was in accord with the rationalistic approach of science. The philosopher could speak of life and consciousness as emergent, but science aimed at describing a universe from which they arose "inevitably out of that which preceded them." (1932: 11) This did not eliminate emergence, since future presents would initially still display such novelty, but once again this novelty would be rationalized away, and so on without end.

Because of Mead's rejection of the past and future, his system was not concordant with science. A scientific description of a thing normally presupposes its existence in space-time, with time considered as an asymmetrically ordered sequence of events. Mead's rejection of past and future, and his notion of rewriting the past as the result of the emergence of novelty, were such as to make science, as the study of things in space-time, impossible. This, however, was independent of his analysis of emergence as a function of sociality.

Mead attempted to provide a grounding for his unusual concept of emergence in a new theory of the social. Each thing was to be considered in its environmental context, and each perspective on a thing-environment complex was a system. In Mead's sense of the terms, a thing could be part of two or more systems, since it might be the object of many different perspectives. A plant or animal, for example, might be analyzed by the physical scientist in the mechanistic sense as a physico-chemical complex interacting with its energy environment, or by the biologist in the teleological sense as a goal-directed organism interacting with its organic environment. Both mechanism and teleology could be accepted as particular points of view, though neither could be admitted as general and all-encompassing standpoints.

An adequate philosophy must recognize the "sociality" of all things, where sociality referred to the simultaneous presence of a thing in different systems. Sociality, like emergence, was then a pervasive characteristic of the present, and as with Wheeler formed the condition for the appearance of the novel. Emergence was explained as the presence of a thing in more than one system, which resulted in properties in any one system which were novel from the perspective of the other systems. Mead distinguished three fields of emergent reality, depending on the existence of these different perspectives on the same thing: the physical, the biological, and the mental. These corresponded to levels of reality in more traditional systems of emergence.

(e) Cosmic Evolution: J. E. Boodin

John Elof Boodin (1869-1950) published his *Cosmic Evolution: Outlines of Cosmic Idealism* in 1925. The point of departure for Boodin was that the universe was divided into levels, of which he mentioned four: the inorganic, the organic, the mental, and the spiritual. This level structure was characteristic of the cosmos as a whole, and in each part of the cosmos there was an evolutionary process whereby these levels were progressively attained. It was through the interaction of the part with the whole – the earth with the cosmos – that this transition from level to level occurred locally: "Since in the universe as a whole all the levels of reality may be supposed eternally to coexist, there would thus be provided the rationale for the evolution in one part of the cosmos from a lower to a higher level of existence without introducing magic." (1925: 36) By "magic" Boodin referred to the concept of emergence as locally generated rather than globally induced.

The higher levels of reality pre-existed as guiding causes: "[T]he actual is prior to the potential and furnishes the plus factor which makes a given level of development potential of a higher one. The stream of evolution does not rise higher than its source." (1925: 78) Boodin believed that traditional emergence was based on the fallacy that new forms and characters could be accounted for in terms of simpler antecedents within the series. In Boodin's view, the controlling field rather than the preceding conditions was the true cause.

Boodin saw three major advantages of his system. Firstly, it avoided the excesses of both materialism and idealism, while satisfying the legitimate desires of each. In accord with materialism, life emerged from matter, and in accord with idealism, the spiritual level of values was the highest level and guiding force. Secondly, it combined epigenesis and preformationism, though each on its own was incomplete. When a part of the cosmos was examined in its relation to the whole, the relation was a preformationist one, that of the local recapitulating the global. But considered in its local context, evolution manifested epigenetic characters, since new levels arose which were not strictly contained in the previous local ones. Thirdly, it incorporated emergentism but went beyond it by explaining emergence as the result of cosmic interaction: "The claims of the emergence theory, so far as it is truly descriptive, are recognized; *viz* there is creative synthesis and emergence of properties, forms, [and] levels, but a rationale is furnished for this emergence in the conception of the interaction of the particular history with the structure of the cosmos." (1925: 128) However, Boodin did not propose any physical mechanism that explained how the whole communicated with its parts in order to induce local emergence.

(f) Holistic Evolution: J. C. Smuts

Evolutionary holism was developed by J. C. Smuts (1870-1950), a South African military officer and political leader turned philosopher in *Holism and Evolution* (1926). Writing at a time when the gestalt psychologists were arguing that what counted was the overall configuration of a complex rather than the structure of its parts, Smuts argued that the formation of "wholes" was the *vera causa* or motive force of evolution. Evolution was a creative process, not a mere unfolding of immanent potentialities, and it passed through the levels of matter, life, and mind to the fourth and final level of Smuts' system, personality.

Matter existed as atoms and molecules, and which in the colloidal state prepared the way for the appearance of cells and organisms. Life, the second level of the system, was based on the cell, which differed from atomic and molecular organization in its far greater complexity of structure and function, increased differentiation of parts, and improved integration as a whole. Life evolved from single cells to multi-celled organisms in which there was a system of organic regulation that coordinated the various parts, and which foreshadowed the subsequent development of mind. Mind, the third level of the system, was incomplete so long as it had not developed into personality. Personality was the highest whole produced by evolution, the outcome of mind endowed with purpose and expressing itself in ideals and values. Mind proceeded from the realm of necessity, where it was intimately linked to body, to the realm of freedom, which was a conceptual domain.

Smuts mentioned Bergson as an inspiration for his theory of holistic evolution, though as a monist Smuts rejected the *élan vital*. He differed with Lloyd Morgan in that he considered the formation of wholes as more fundamental than the emergence of novelty, which was analyzed as a consequence, not a cause. Further, while Lloyd Morgan accepted psycho-physical correlation or at least psycho-physiological concomitance, Smuts admitted the mental as a level achieved by evolution after that of life, and not co-extensive with it. He noted that Lloyd Morgan's theory "seems to be a reversion to the preformation type of evolution and to be destructive of all real effective 'emergence'." (1926: 321) Finally, while Lloyd

Morgan accepted a role for deity in his system, Smuts avoided the theological question entirely. Smuts' system bore some similarity to that of Sellars, who also recognized four levels in a naturalistic universe, with the difference that the fourth level was society for Sellars and personality for Smuts.

(g) Levels and Realms: Nicolai Hartmann

Nicolai Hartmann (1882-1950) incorporated emergentist ideas in *New Ways of Ontology* (1953), a translation of *Neue Wege der Ontologie* (1949), which was based on work he had begun in *Der Aufbau der Realen Welt* (1940). His goal was to develop an ontology which combined being and becoming as an integral unity, rather than treating them as separate concepts, externally and contingently related: "Becoming is no opposite of being but is a form of being" and ontology must address itself to the "Being of Becoming." (1953: 28) His ground for ontology was a system of categories, considered not as Kantian categories of the mind, but as "universal constitutive principles" (1953: 60) which were discovered empirically, not deduced *a priori*.

Though Hartmann's initial division of reality into the two realms of the spatially extended material and the non-spatial mental was similar to Cartesian dualism, there was an emergentist component when he considered the strata into which the two realms were sub-divided. The material realm was subdivided into the physical and the biological strata, while the mental realm was sub-divided into the psychological and the spiritual strata, this latter a level of culture, not of deity.

The categories played an essential role in the dynamic construction of the strata of reality. There were two types of categories: those that recurred at all levels, of which Hartmann mentioned only three – time, process, and causality – and those that occurred only at specific strata, such as space, which Hartmann limited to the material realm of the physical and the biological. Looked at from the dynamic point of view of becoming, higher strata emerged and new ontological "heights" occurred with "the emergence of categorial novelty." (1953: 76) As in other emergentist analyses there was both continuity and novelty: "Recurrence [of categories] guarantees the continuity, novelty [of categories], the diversity of strata." (1953: 81) However, Hartmann's system differed from traditional emergentism in neglecting the role of evolution, and in the quasi-Cartesian distinction of realms.

On the basis of the above analysis of the four strata and their associated categories, Hartmann criticized other philosophical systems which universalized the categories of one level at the expense of those of the others, and so neglected the emergence of new categories and their associated levels. Materialism was criticized as "metaphysics from below," mistakenly elaborating the characteristics of higher levels deductively from those of the lowest level. Idealism was the converse mistake of "metaphysics from above," and attempted to deduce lower level concepts such as matter from higher level ones such as mind or idea. Biologism and psychologism narrowly based themselves on one of the intermediate levels to the exclusion of those above and below them. Hartmann characterized his position as one combining multiplicity and unity, two of the categories: multiplicity of levels, unity of system.

(h) Levels and Branchings: James Feibleman

James K. Feibleman (Tulane University), published his *Ontology* in 1951, though he indicated in the foreword that he had begun work on it in 1945, and had published sections in article form between those two dates. Like Hartmann, Feibleman argued that a new ontology was needed, one which would be based on categories sufficient to encompass all aspects of being and becoming. Unlike Hartmann, he held that there must be just three basic categories. One category did not suffice for the multiplicity of reality, which was characterized by being and becoming, while two were inadequate, since there was a need for a third to mediate between them. Four or more were unnecessary, since additional ones would detract from the intermediary role of the third and be redundant.

Feibleman identified the three fundamental categories as those of essence, existence and destiny. The first two were considered as generating distinct universes: essence determined the universe of being as the realm of possibility, and existence determined the universe of becoming as the realm of actuality. Destiny was the relation linking the two and so did not constitute a distinct universe. The universe of essence was complete and, as in Plato, independent of the existence of changing things, while that of existence was made up of changing things which interacted with each other.

The universe of being was divided into two sub-systems, the theoretical and the empirical. The theoretical was further divided into the three domains of the ontological, the logical, and the mathematical, while the empirical was sub-divided into the five domains of the physical, the chemical, the biological, the psychological and the cultural. The series of domains was arranged from lowest to highest as follows: "Infra-physical (if any), Physical, Chemical, Biological, Psychological, Cultural, Mathematical, Logical, Ontological, Ultra-ontological (if any)." (1951: 268)

According to Feibleman, there was an element of discontinuity between any two domains when the series was read from highest to lowest, since in this case the accent was on the distinctness of each higher domain relative to the lower domains. But there was an element of continuity when the series was read from lowest to highest, with the accent on the lower domain preparing the way for the one immediately above it. The basic relation between a higher and lower domain was that the higher was more complex than the lower.

It was with the universe of existence that emergence played a dynamic role. There were three basic processes of change: emergence as coming-into-existence, endurance or duration as remaining-in-existence, and transience as passing-out-of existence. The world of becoming was an active one of change and the flow of energy, not a static one of universals and values as in the universe of essence.

There was a structure of "integrative levels" in this universe starting from the bottom-most level, that of matter and its quantum forces, leading on to the atomic and molecular levels, and branching to the cosmological and the biological. Within this latter, there was the development of the plant, animal, and human levels, including societies, mind, and culture. Feibleman's diagram of the levels of reality in the universe of existence featured bifurcations and was presented as an upward developing tree structure: (1951: 336)

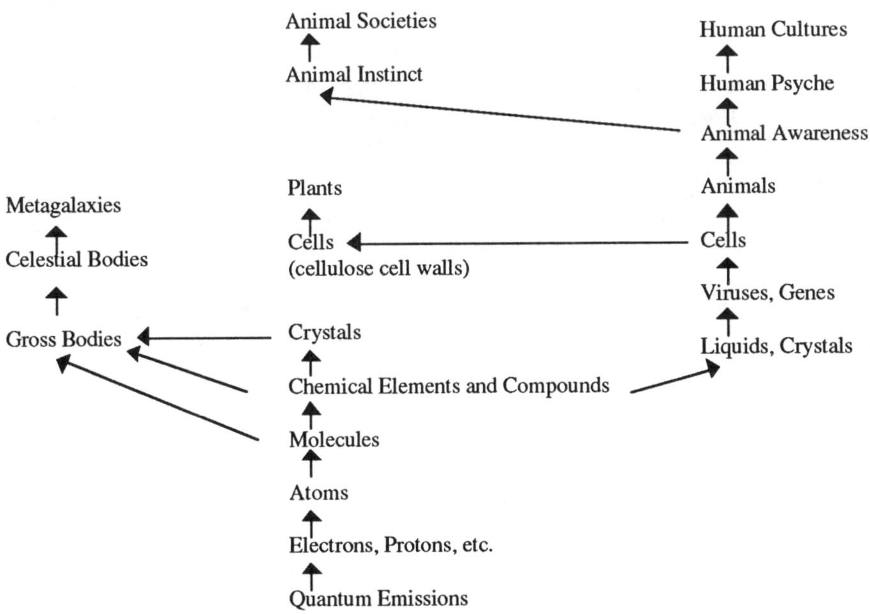

Diagram 11.1: Feibleman's Representation of the Levels of Being

Feibleman further developed his theory of levels in "Theory of Integrative Levels." (1954), where he provided a summary of his theory in twelve points:[2]

> 1. Each level organizes the level or levels below it plus one emergent quality; 2. Complexity of the levels increases upward; 3. In any organization the higher level depends upon the lower; 4. In any organization, the lower level is directed by the higher; 5. For an organization at any given level, its mechanism lies at the level below and its purpose at the level above; 6. A disturbance introduced into an organization at any one level reverberates at all the levels it covers; 7. The time required for a change in organization shortens as we ascend the levels; 8. The higher the level, the smaller its population of instances; 9. It is impossible to reduce the higher level to the lower; 10. An organization at any level is a distortion of the level below; 11. Events at any given level affect organizations at other levels; 12. Whatever is affected as an organization has some effect as an organization. (1954: 59-63)

There was at least one major affinity between Feibleman's system and that of Conger because of the inclusion of the logical and mathematical levels preceding the empirical levels in the universe of essence. Feibleman's concept of "integrative levels" was influenced by the use of that term among biologists during the period of the 1920s and 1930s, though he was original in his inclusion of branchings in his structure of the levels of reality in the universe of existence. Feibleman, like Hartmann, attempted to integrate emergentism within a metaphysical system which belonged to a different trend of thought – Platonic essentialism in his case, and Cartesian dualism in the case of Hartmann.

Other philosophers discussed in this chapter also accepted emergence, not as a central concept, but rather as one derived from more basic ones: Conger's epitomization, Reiser's stochastic processes, Montague's dual factors, Mead's social present, Boodin's local recapitulations, and Smut's wholes. The common strength of all these varied approaches was that they rejected the "natural piety" which Lloyd Morgan, Alexander, and even Sellars endorsed, and which had put an end to further analysis of the concept of emergence. Though there were weaknesses in their treatment that have been identified in this chapter, nonetheless, their efforts were further evidence of the status of emergentism as a trend of thought with both adherents who placed the concept at the center of their work, and sympathizers who incorporated it as part of systems with different central notions.

CHAPTER TWELVE

Critical Reaction to Emergent Evolution

At its apogee, emergent evolution found its finest medium of expression in metaphysical systems such as those of Lloyd Morgan and Samuel Alexander. As a philosophical trend it was at its height during the period of the mid-1920s. A session devoted to it at the VIth World Congress of Philosophy, as well as a colloquium held at the Aristotelian Society were indicative of the interest for the new concept.[1] As the trend of emergent evolution developed, a number of criticisms were made. Stephen Pepper rejected emergentism as an incoherent concept, while Charles Baylis argued that emergence was so ubiquitous as to be trivial. William McDougall subjected emergent evolution to a general critique from the point of view of interactive dualism, while Bertrand Russell rejected it from the point of view of neutral monism. Finally, Rudolf Carnap in his defence of reductionism provided the alternative to emergence that became dominant during the period of the mid-1930s through to the early-1950s during what I term the "eclipse of emergentism."

(a) Emergence as Epiphenomenal or non-Causal: Stephen Pepper

Stephen C. Pepper in "Emergence" (1926) argued that emergent qualities were epiphenomenal and useless, while emergent laws were redundant and expendable. In his argument against emergent laws, Pepper considered as typical the case where four variables q, r, s, and t at level B were related by some law $f_1(q, r, s, t)$. He supposed that r and s gave rise to a new emergent quality at a further level C, and that a new law $f_2(q, r, s, t)$ described the relationship at that new level. Pepper then argued that this could not be the case, as either f_1 sufficed or f_2 did, but both could not be distinct and yet describe the same situation: "The point is, either f_1 adequately describes the interrelationships of (q, r, s, t) or f_2 does; or if neither adequately describes the interrelationships there is some f_3 that does, but there can not be two adequate descriptions of the same inter-relationships among the same variables." (1926: 242)

Pepper's argumentation, however, was based on the recurrence of r and s as variables in laws at both levels B and C, despite the fact that a new emergent system c had arisen at level C from components r and s at level B. On an emergentist view, the law at level C, using Pepper's formalism, should be $f_2(q, c, t)$, where $c = f_4(r, s)$ and not $f_2(q, r, s, t)$ as Pepper stated. In this case, f_2 and f_1 are different, and Pepper's objection would not hold.

Pepper further claimed that emergent laws contradicted the "aim of science" which he defined as the identification of a small number of basic laws from which all phenomena could be deduced, following the axiomatic model of mathematics. But mathematics is a formal science, while physics, chemistry, biology, and psychology are factual ones, and it is not *prima facie* evident that an ideal for the formal sciences could be transferred automatically to the factual ones. Moreover, it is questionable whether this is an ideal at all even for mathematics, given the failure of logicism to show that all of mathematics can be derived from a simple axiomatic basis.

Pepper's claim that emergent qualities, as distinct from emergent laws, were epiphenomenal can be reformulated as follows: it makes no difference for entities at the lower B-level that entities at a higher C-level have emerged. But the simple case of the emergence of plants at the C-level of life does make a difference for chemical compounds at the B-level. The plant, made up of DNA-based cells, has the capacity or quality of reproduction. In the course of its life-cycle it will take up nutrients from the soil, and this will effect the composition of the soil, depleting it of those types of substances the plant uses as food. The emergent quality of reproduction at the C-level therefore makes a difference to the composition of the soil at the B-level and so the emergent quality of reproduction is not epiphenomenal.

(b) Emergence as Ubiquitous and Trivial: Charles Baylis

The criticism by Charles Baylis in "The Philosophic Functions of Emergence" (1929) was more profound. Rather than admitting only integrative emergence, where a new property appeared as the result of the integration of parts into a whole, Baylis considered three further possibilities: integrative submergence, where some properties of the parts were lost in the resulting whole, disintegrative emergence, where new properties arose as the result of the separation of a whole into its parts, and disintegrative submergence, where properties of the whole were lost in the process of its decomposition into parts. Baylis' claim that all four types were legitimate occurrences can be summarized in tabular form as follows:

type	*integration*	*disintegration*
emergence	integrative emergence	disintegrative emergence
submergence	integrative submergence	disintegrative submergence

Table 12.1: Baylis on Emergence and Submergence

Baylis discussed these four possibilities using the standard case of the combination of hydrogen and oxygen to form water. Integrative emergence occurred when at normal temperature and pressure water was a fluid, though under the same conditions neither of its components had that property. Integrative submergence occurred when water lost the property of being a gas which both of its components had at normal temperature and pressure. Disintegrative emergence occurred when hydrogen and oxygen were separated from water and exhibited the property of being a gas which they did not possess as members of the compound. And finally, disintegrative submergence occurred when hydrogen and oxygen were separated from water and lost the property of being part of a liquid substance.

Baylis concluded that emergence and submergence were phenomena associated with every change: "Emergence and submergence, integrational and disintegrational, occur in the realms of physical, mental and logical entities, not merely occasionally but ubiquitously." (1929: 375) It was precisely their ubiquity which ruled them out as the basis for a new philosophy, since such a philosophy would be just a terminological variant of the traditional philosophy of change and becoming.

Moreover, Baylis believed that final assurance could never be achieved that a property was emergent. By the ubiquity argument, every complex had at least one emergent or submergent property. But which one? For *X* to be an emergent property of a complex *R(A,B)*, it must not be a property of *A* or *B* or any proper part of *A* or *B*. But analysis was never complete, and there could be no guarantee that *X* was not a resultant property of some part at a further level of decomposition. Baylis concluded: "These two facts, the ubiquity of emergence, and the difficulty of ever being sure that a particular character is an emergent of a particular complex, render the concept almost useless philosophically." (1929: 377) As an afterthought, Baylis added the criticism that the dispute among emergentists as to the number and types of levels of reality was a failing of the theory: "Such disagreement itself indicates the futility of the labor. The truth seems to be that any number of levels may be picked out according to the taste and inclination of the chooser." (1929: 378-39)

Baylis' criticisms were more important and better formulated than those of Pepper, as he introduced the important distinction between emergence and submergence, and their relation to the integration and the disintegration of wholes. Baylis' argument rested to a large extent on his contention that every change, or at least every change bringing about an integration of components or a disintegration of a whole, involved emergence or its converse, submergence, so that the emergence-submergence couple became a mere synonym for change. But in the cases of the emergence of life, society, and mind, significant changes occur – the appearance of new levels of reality – which do not take place in other types of simpler integration, such as the interaction of species in an ecosystem, or in changes which usually do not produce qualitative novelty at all – such as change of place or orientation. Emergence of qualitative novelty productive of new levels of reality is a proper subset of changes in reality, and so is not ubiquitous. Moreover, the debate over the number and nature of the levels of reality was not a weakness of emergentism as a system, but a sign of its vitality as a philosophical trend that engendered healthy disagreement and controversy.

(c) Dualistic critique of Emergentism: William McDougall

The foremost critic of emergentism during the 1920s was the psychologist William McDougall (1871-1938), whose *Modern Materialism and Emergent Materialism* (1929) was the only book length criticism of emergent evolution published in England. McDougall had earlier adopted an interactionist and dualist approach to the mind-body problem. In "On the Seat of the Psycho-Physical Processes" (1901) he argued, following Hermann Lotze, that there was no one "seat of the mind," such as the pineal gland in Descartes' theory. Rather, there were many sites of mind-brain interaction, which McDougall identified with the synapses of the neurons: "And if the view that I advocate should prove to be true, then those who adopt the spiritualistic hypothesis will have to regard the synapses as the places of interaction of the body and soul." (1901: 580) McDougall combined this thesis with a correspondence view of the relation between ideas, considered as mental objects, and complexes of neurons, the material basis for ideas. In terms anticipating D. O. Hebb's cell assembly theory, though not his materialist ontology, McDougall stated that to each idea there corresponded a complex of linked neurons, and that a compound idea resulted when two neuronal complexes were united through the facilitation of nerve impulses between them.

In *Physiological Psychology* (1905), McDougall proposed an analogy between mind-brain interaction and electromagnetic fields which can be considered as proto-emergentist. When an electric current passed through a wire, a magnetic field was also established about it. When two such wires were placed near each other, a complex interaction resulted, with mutual modifications of the electric currents and magnetic fields until an equilibrium was achieved. McDougall made the following correspondence between the electromagnetic interaction and the mind-body problem: "In this crude simile the wires stand for nervous arcs, the electric current for the flow of nervous energy through each arc, the magnetic field generated by the current in each wire flowing separately for a psychical element, and the total magnetic field, when several or more wires are in action, for the state of consciousness." (1905: 168) Just as the total magnetic field was different from the sum of the fields generated by electrical flows separately in each wire, so too the state of consciousness was more than the mere sum of the individual psychical elements.

It was undoubtedly this model of mind-brain interaction which McDougall referred to in his "Autobiography" (1930) as anticipating emergentist and holistic views of subsequent authors: "I continued to hold the view ... that the psychical qualities are engendered by (or as would now be said "emerge from") the complex conjunctions of brain-processes (now called "configurations") not as mere epiphenomena, but rather as synthetic wholes that react upon the physical events of the brain or have causal efficacy within the whole complex of psycho-physical events." (1930: 205-206)

The electromagnetic model would indeed be fully suggestive of an emergentist theory were it not for McDougall's further extension of the model in a dualistic direction. McDougall, writing in the year of Einstein's special theory of relativity but unaware of it, made a correspondence between the aether, as the supposed immaterial medium for electromagnetic fields, and the soul as the immaterial ground of mind. But whereas science rejected the immaterial aether as the result of Einstein's derivation of relativity theory from electromagnetic considerations, McDougall went in the opposite direction in his philosophy. He maintained the psychic element of immaterial mind which he had metaphorically compared to the aether, and discarded the electromagnetic analogy. In so doing, he recognized the incompatibility of the two positions: either the mental emerges from complex interactions of neuronal systems, or the mental is the result of an immaterial psychical element, but not both.

As emergent evolution developed as an alternative to mechanism, McDougall turned his critical attention to it. Since he had already speculated on a proto-emergentist theory of the mind-brain relation in the early 1900s and rejected it for a full fledged dualism, he was well prepared for the attack, which was made in *Modern Materialism and Emergent Evolution* (1929). McDougall was concerned, as the title indicated, not only with the problem of emergence in evolution, but also with what he considered to be its materialist ontology. McDougall argued that emergence was intimately related to materialism because most emergentist systems started with matter as the first level.[2]

McDougall's repugnance for materialism was so great that although August Weismann had refuted Darwin's theory of pangenesis as a material basis for Lamarckian use-inheritance, and thrown the whole concept into disrepute, McDougall persisted in his belief that use-inheritance, despite its lack of a material basis, did in fact occur at the mental level. In "Mental Evolution." (1925), he argued that since the inheritance of acquired mental characteristics did not have a material basis, it must have an immaterial basis: "If, then, Lamarckian transmission occurs, it is in itself good evidence of the reality of that immaterial organization of the facts of memory, the facts of the unity of consciousness, the facts of integration and disintegration of personality, the facts of intelligent purposive activity." (1925: 155) A more reasonable conclusion would have been to reject Lamarckian use-inheritance, based on the following counter-argument: if there were no material basis upon which inheritance of acquired mental characteristics could be based, then the phenomenon itself would be impossible, and reports of it must be regarded with suspicion and be rejected. McDougall, however, preferred to defend the much criticized and dubious phenomenon of use-inheritance in order to serve his own goal of defending mind-body dualism. And his defence of dualism motivated and informed his rejection of emergentism.

Consequent with his dualism, McDougall held that mind did not emerge from matter, but had its source in an immaterial psychic principle. Once mind arose, it developed through a process of differentiation in which there was again no emergence of new mental faculties: "There has been, not evolution of Mind from the physical realm, but evolution of mental capacities – and this evolution has been characterized by a progressive differentiation of the powers of Mind, rather than by emergence of new kinds of relation, causal or other." (1925: 157) McDougall did, however, believe that some products of mental activity had qualities that their parts did not possess. A fugue was more than the sum of the notes that composed it, to use one of his examples, and here, rather belatedly, was an example of an emergent property.

(d) Monistic and Neutralist Critique: Bertrand Russell

Bertrand Russell's (1872-1970) great contribution to modern philosophy was the analytical method featuring the use of symbolic logic to formalize and analyze philosophical problems. Russell based his metaphysics on science, both the formal ones of mathematics and logic, and the factual ones of physics and psychology. His philosophy of neutral monism held that events, which he took to be basic, could be organized equally well by either physics or psychology. As organized by psychology, they were mental, while as organized by physics, they were material. Though he recognized physics and psychology as essential and parallel sciences, Russell accorded no equivalent place to biology in his scientific world view. This omission of the biological played a major role in his rejection of the philosophy of evolution, though in the mid-1920s, he would briefly consider emergentism as an alternative to his own neutral monist ontology.

In *Our Knowledge of the External World* (1914), delivered as the Lowell lectures for that year, Russell associated the evolutionary philosophies of Spencer and Bergson with progress and teleology, two concepts he considered out of place in a scientific philosophy, since they were merely projections of human desires rather than expressions of scientific fact. However, although Russell was highly critical of evolutionary philosophy, he did not reject biological evolution as a real process in nature. But as a process perhaps unique to one planet, it could not serve as the basis for a general philosophy.

Given this negative view of evolutionary philosophy, it is of interest that Russell seriously examined emergentism in *Philosophy* (1927) as the result of reading Broad's *Mind and Its Place in Nature* which had just appeared the year before. Russell noted that "the possibility [of emergence] is an important one, and it will be worth while to consider it." (1927a: 293-294) He proceeded to enumerate the following propositions: matter may indeed be emergent from events, at least as far as its quantum mechanical properties are concerned; mind may be emergent from events, at least insofar as the qualitative aspects of sensation are concerned; but mind is not emergent from matter since the mental series runs parallel to the physical one, ruling out emergent evolution as a metaphysical solution to the mind-body problem. Neutral monism, *pace* Lloyd Morgan, was considered as an alternative to emergent evolution, and not as a complement to it.

In his *Analysis of Matter* (1927), Russell devoted just one paragraph to emergentism. In discussing the relation of events to things, and defending the legitimacy of analysis, he stated: "We may say that this is the characteristic merit of analysis as practiced in science: it enables us to arrive at a structure such that the properties of the complex can be inferred from those of the parts." (1927b: 285-286) Emergence was at best an epistemological category applicable in cases where properties had not yet been fully analyzed by science:

> Dr. C. D. Broad, in the *Mind and Its Place in Nature*, lays stress upon what he calls "emergent" properties of complexes – i.e. such as cannot be inferred from the properties and relations of the parts. I believe that "emergent" properties represent merely scientific incompleteness, which would not exist in the ideal physics. It is difficult to advance any conclusive argument on either side as to the ultimate character of apparently "emergent" properties, but I think my view is supported by such examples of the explanation of chemistry in terms of physics by means of the Rutherford-Bohr theory of atomic structure. (1927b: 286)

Russell's opinion of emergentism continued to diminish, and in *The Scientific Outlook* (1931) he discussed it as a form of "evolutionary theology." Russell criticized as apologists those who saw in evolution "evidence of a Divine Plan slowly unfolding through the ages," (1931: 125) since the very existence of evil in the world negated the thesis that it was under a benevolent and omnipotent governance. Russell focused on a quote from Lloyd Morgan to the effect that emergent evolution was "from first to last a revelation of that which I speak of as Divine Purpose." (in 1931: 135) Russell erroneously considered this as either a definition of emergent evolution, or at the very least, an essential characteristic of it. But for Lloyd Morgan, the thesis of *nisus* towards deity was not the core of his theory, which was instead furnished by the theses concerning the emergence of life and mind as levels of reality. The teleological-theological element was only a part of some systems of emergent evolution, and it could be rejected, as Sellars did, without sacrificing the integrity of the philosophy. Russell reasoned as follows: divine purpose is essential to emergent evolution; divine purpose is to be rejected; therefore, so is emergent evolution. The argument is valid, but since the major premiss is false, there is no warrant to conclude to the truth of the conclusion.

(e) The Eclipse of Emergence: Carnap and Reductionism

The period of the mid-1930s to the mid-1950s marked the "eclipse" of emergentism, using this concept on analogy to that of Peter Bowler in *The Eclipse of Darwinism: Anti-Darwinian Evolution Theories in the Decades Around 1900* (1983). Bowler argued that weaknesses in Darwinian theory and advances in Mendelian genetics around the turn of the century resulted in the theory of evolution by natural selection being temporarily supplanted by the theory of evolution by mutations. Yet Darwinism, or at any rate, neo-Darwinism, rebounded in the modern synthesis during the decade 1935-45, and became once more part of the mainstream view of evolution. The concept of "eclipse" refers to a temporary overshadowing of one theory by another, and this was certainly also the case when emergentism, a non-reductive philosophy of biology and psychology, was temporarily overshadowed by reductionism and the physicalist thesis that all the sciences could be reduced to physics. Despite a number of articles on emergence in philosophical journals, no new systems of emergent evolution were produced in book form during this period from the mid-1930s to the early 1950s.[3] The work of Feibleman and Hartmann came at the end of this period, and admixed elements of Platonism and Cartesianism with emergentism.

The reasons for the eclipse of emergentism were multiple. Lloyd Morgan had retired from teaching in the late 1910s, and though he maintained emeritus status, he had no students to continue his work after he systematically expounded the theory of emergent evolution in his Gifford Lectures of 1922-23. Broad switched his interest from emergence to the defense of parapsychology and the paranormal, Wheeler concentrated his efforts on the scientific investigation of insect societies, while Alexander worked on aesthetics during the last years of his life. Though Roy Wood Sellars continued to discuss emergentist ideas in his articles, the books he wrote during the rest of his long life were all devoted to critical realism in epistemology, with the exception of a second edition of *The Principles and Problems of Philosophy*, which was issued in 1970 – 45 years after the first edition – as *The Principles, Perspectives and Problems of Philosophy.*

System-building, especially within the Anglo-American world, was suspected of speculative excess as interest shifted to more focused and narrower problems susceptible of investigation using the modern techniques of logic. Works such as Lloyd Morgan's *Emergent Evolution*, and Alexander's *Space, Time and Deity* appeared to be the last gasp of older style of speculation, to be replaced by the newer methodology of analysis. Within philosophy of science, emergent evolution, never formulated in symbolic and exact terms, and concerned mainly with the problems of life and mind, could be dismissed as irrelevant as philosophical attention centered on the new problems of quantum mechanics and the foundations of physics. Physicalism proposed the goal of reducing all of science to the base level of physics, a project systematically undertaken by Rudolf Carnap.

Carnap (1891-1970) believed that philosophy had to be reformed in order to bring it in line with the development of the new symbolic logic of Russell and Whitehead, further refined by David Hilbert and others. In *The Logical Structure of the World* (1928) he opposed system building, arguing that philosophy could no longer continue as an activity centered on the speculations of individuals concerning the metaphysical nature of reality. It had to become a cooperative endeavor, based on each philosopher's contribution to the common project of clarifying the epistemological foundations of science. Carnap's philosophy involved an essential leveling down, so that chemistry, biology, and psychology would ultimately be reduced to physics. Along with this reduction to the physical was the rejection of traditional metaphysical problems as non-scientific and incapable of resolution.

In considering the case of the mind-body problem, which emergentists believed to have addressed philosophically and solved in a new way, Carnap sharply distinguished between the scientific aspect of the problem, which aimed to discover and describe the relations between brain states and mental states, and the metaphysical aspect, which tried to explain these correlations based on the concept of substance. On Carnap's view, the scientific problem was well-founded and capable of resolution, while the metaphysical problem was not because it invoked non-empirical concepts like "substance." A distinction was made between "correlational" problems which science could address, such as the determination of the brain centers associated with emotion, and "compositional" problems which lay outside of science, such as whether the nature of mind was material or spiritual. Carnap suggested that metaphysical questions like the mind-body problem belonged more to poetry than to philosophy. The immediate object of Carnap's critique was German speculative thought, Heidegger's "bad poetry" in particular, and Carnap did not explicitly criticize emergentist ideas. When the question as to his attitude towards emergence was later posed to him by Herbert Feigl in the *Library of Living Philosophers* volume (Schilpp, 1963) devoted to Carnap's work, he not unexpectedly replied in the negative, a point to be further explored in the following chapter.

Whereas emergentism argued for a metaphysical system based on science, the logical positivism of Carnap called for the elimination of metaphysics in favour of an epistemological foundation for science. Whereas emergentists only occasionally referred to logic, Carnap's philosophy was based upon it. And whereas the emergentists were largely concerned with biology and its relation to psychology, including the mind-body problem, Carnap considered this latter question nonsensical, and turned his attention to physics, and its construction from an inter-subjective basis of empirical concepts and verifiable propositions. The gap between emergentism and reductionism was immense, and so long as reductionism dominated philosophy of science, emergentism was eclipsed.

CHAPTER THIRTEEN

Further Discussion of Emergent Evolution

However, reductionism, even at its apogee, never completely supplanted emergentism in one area – that of the philosophy of biology – where proponents of the theory of "integrative levels" defended a hierarchy of organic units featuring novel properties, and recognized that though biology was based on the physical sciences, it was not reducible to them. The reductionist project, even if it resulted in achievements such as the *Encyclopedia of Unified Science*, did not succeed in reducing all of science to physics, a point which the logical positivists themselves came to admit.[1] Since the mid-1950s, the concept of emergentism has been rehabilitated, to the point that it has once more taken its place as a legitimate basis for philosophy of science and metaphysical systems based on science.

(a) Integrative Levels: from Joseph Needham to Alex Novikoff

Joseph Needham, an embryologist and author of the multi-volume *Science and Civilization in China*, was also a prominent exponent of the concept of integrative levels. In his 1937 Spencer Lecture, "Integrative Levels: A Re-evaluation of the Idea of Progress," he defined his theme as "the existence of levels of organization in the universe, successive forms of order in a scale of complexity and organization." (1937: 234) Each level of organization was characterized by a specific form of evolution, and Needham mentioned the cosmological, biological, and sociological, with mental development included in the latter. At each new level there was greater complexity of structure, efficiency and centralization of control, and autonomy of systems.

Needham recognized Lloyd Morgan and Sellars as authors of similar points of view, and added: "I am not quite sure where the term "levels" was first used in this way, perhaps in S. Alexander's *Space, Time and Deity*."[2] (1937: 239) At the same time, Needham associated the concept of integrative levels with the organicist conception of nature of Alfred North Whitehead. Organicism was taken to be the view that biological structures could not be fully comprehended in isolation, but must be considered in their environmental context, which for tissues and organs was the organism itself.[3]

In the 1930s Needham became a Marxist and combined elements of organicism, dialectics, and emergentism in his philosophy of biology. In the Terry Lectures of 1936, *Order and Life*, he stated "that biological order is a form of order different from those found in physics, chemistry, or crystallography, yet not impenetrable by the human mind or ruled by unintelligible spiritual entities. Translated into terms of Marxist philosophy, it is a new dialectical level." (1936: 45) A dialectical materialist concept of irreducible levels was also introduced in the second edition of A. I. Oparin's *The Origin of Life* (1938), where Oparin argued that though life arose naturally from a physical-chemical basis, that of colloidal structures in particular, it also had a specificity which could be dealt with only in biological terms.[4]

A major reinforcing influence on Needham's conception of the level structure of reality was J. H. Woodger (University of London), whose *Biological Principles: A Critical Study* appeared in 1929. Woodger argued that the concept of organization was key to a new philosophy of life which would overcome the old antagonisms between vitalism and mechanism, a theme also developed by C. D. Broad. Organization essentially involved a hierarchical part-whole relationship, with distinct levels of organization that could not be reduced one to the other. Woodger, who was a collaborator of Ludwig von Bertalanffy, went on to develop this theory in a series of three articles which appeared under the title "The 'Concept of Organism' and the Relation Between Embryology and Genetics."[5] (1930-31)

Though Woodger was only secondarily interested in emergence and focused mainly on the concept of levels, H. S. Jennings (Johns Hopkins University) was more explicit in his endorsement of emergent evolution. In a lead article in *Science* for Jan. 14, 1927, which reproduced his address as retiring chairman of the Zoological Section of the American Association for the Advancement of Science, he stated: "The doctrine of emergent evolution is the Declaration of Independence for biological science," (1927: 22) since it specified the biological as a distinct and irreducible level of reality.

This enthusiasm was not unanimously shared, and the geneticist T. H. Morgan (1866-1945), in his Messenger Lectures of 1931, *The Scientific Basis of Evolution* (1932) criticized the concept. As the foremost experimentalist in the new discipline of genetics, Morgan had developed the concept of gene mutations as the source of evolutionary change, replacing the continuous morphological variations of Darwin. But though he disagreed with Darwin on the mechanism of evolutionary change, he agreed with him that the mode of change was mechanistic. Commenting on the familiar example of water as an emergent, Morgan noted that Mendeleyev in his periodic table had successfully predicted the chemical properties of missing elements based on those of existing and known elements. This implied that such a reduction would eventually also be possible for the properties of chemical compounds relative to the properties of their atomic components. Morgan believed that the biological, too, would be reduced to the inorganic sciences: "Modern biology, then, rests on the assumption, sometimes amounting to a conviction as the result of wide experience, that the properties of living things are the outcome of their chemical and physical composition and configuration." (1932: 236). He continued that this did not amount to saying that the "known laws" of the inorganic were sufficient to explain the biological, but his preference was for a mechanistic explanation based exclusively on further physico-chemical laws, some of which might be discovered by investigating living systems, rather than for an emergentist view which he considered as metaphysical and out of place in science.

Julian Huxley (1887-1975), the biologist who coined the phrase "the modern synthesis" to designate the merger of Darwinian evolution and Mendelian genetics, defended an emergentist philosophy of levels and values in his Romanes Lecture of 1943. This lecture bore the same title, "Evolution and Ethics," as that given by his grandfather T. H. Huxley exactly half a century before, and was intended to improve on the latter's negative evaluation of evolutionary ethics. Julian Huxley defined evolution as a "comprehensive and continuous process from star-dust to human society." (1947: 120) Evolution was creative of novelty, in the sense that "now and again there is a sudden rapid passage to a totally new and more comprehensive type of order of organization, with quite new emergent properties, and involving quite new methods of further evolution." (1947: 120)

Fifty years previously, T. H. Huxley, operating under reductionist assumptions, had been unable to accept evolutionary ethics, since ethics required altruism and cooperation among individuals, and this was inconsistent with the struggle for survival and egoism he believed to be essential to evolution. However, on an emergentist view, Julian Huxley argued that values developed as society evolved, and new qualities developed that were not mere reflections of biological processes. He concluded that an emergent evolutionary process opened up possibilities for new forms of organization and behaviour, including cooperative activities that could be a basis for ethics. More complex societies allowed for greater respect for individual differences, and promoted measures to minimize the distress, discomfort, and suffering of others: "In the broadest possible terms, evolutionary ethics must be based on a combination of a few main principles: that it is right to realize ever new possibilities in evolution, notably those that are valued for their own sake; that it is right both to respect human individuality and to encourage its further development; that it is right to construct a mechanism for further social evolution which shall satisfy these prior conditions as fully, efficiently, and rapidly as possible." (1947: 145) Because of his adoption of emergentism, Huxley was also interested in Teilhard de Chardin, whose philosophy incorporated emergent levels of reality in an evolutionary system, though its goal, unlike Huxley's, was mystical union with the divine.[6]

The theme of integrative levels was the subject of a symposium organized in 1941 to mark the 50th anniversary celebration of the University of Chicago, bringing together biologists and sociologists in proceedings that were edited by R. Redfield and issued as *Levels of Integration in Biological and Social Systems* (1942). R. W. Gerard (Dept. of Physiology, Univ. of Chicago) in "Higher Levels of Integration" dealt with the relation between the biological and the social, a topic which had first been investigated in an emergentist context by W. M. Wheeler. Gerard identified three integrative levels of organization which he termed "orgs," "animorgs," and "epiorgs." He called any organized entity consisting of parts an "org," whether it was animate or not. Advancing integration in the series of orgs resulted in the formation of hierarchical structures of orgs consisting of orgs, such as molecules composed of atoms. Orgs which manifested the characteristics of life – metabolism, growth, and reproduction – were termed "animorgs." Societies were analyzed as "epiorgs" – animorgs organized by social relations. Since epiorgs functioned like animorgs, the social group could be regarded as "a single biological unit." (1942: 75) The related notion of societies as "supraorganisms" was developed by A. E. Emerson in "Basic Comparisons of Human and Insect Societies" (1942), his contribution to the Chicago symposium.

This led to a polemic with Alex B. Novikoff (Dept. of Biology, Brooklyn University), who submitted an article to *Science* entitled "The Concept of Integrative Levels and Biology" (1945). Novikoff defended a four level structure of reality, enumerated as the physical, chemical, biological, and sociological. These were the products of the "evolution of matter through successive and higher stages of integration." (1945: 209) His exposition was based on typical emergentist formulations: new levels arose as a result of growing complexity of organization, each level was dependent on the lower level, but manifested certain unique and novel properties of its own, knowledge of lower levels was a necessary but not a sufficient condition for knowledge of the higher levels, and higher level laws could not be predicted *a priori* from the lower level ones.

Novikoff was particularly critical of the failure to grasp the distinction between the biological and the social levels. He rejected those theories which extended the social to all biological species as wrongly endowing lower life forms with higher attributes, and he rejected as committing the converse mistake of reducing the social to the biological those theories which considered the social as merely a function of the biological. It was this latter error which Novikoff considered Gerard had made: just as an organism was not a society, so societies were not organisms.

Joseph Needham, then in China, sent in a note to *Science* commenting on Novikoff's article. Though he applauded Novikoff's emphasis on integrative levels and complexity of organization, he supported Gerard's organicism as well. Needham believed that Novikoff had overestimated the break between animal and human societies, and too sharply drawn a line of demarcation between the biological and the social. Gerard and Emerson replied in a longer article "Extrapolation from the Biological to the Social" (1945), in the same issue of *Science*. Their claim was that societies were "living systems that obey those most general laws which apply to all living systems." (1945: 583) Though new and unexpected properties emerged at successive levels of organization, "the new properties must be commensurate with the old; must fit in their general framework, not violate it." (1945: 584) In his characterization of societies as epiorgs, Gerard had taken a reductive position that followed from his organicist assumptions – all entities are organisms of a sort, – and which Novikoff, arguing from an anti-organicist point of view, had rejected.

Novikoff's influence was evident in *Philosophy for the Future: The Quest of Modern Materialism*, which Roy Wood Sellars edited, along with V. J. McGill and Marvin Farber. C. Judson Herrick (1868-1960), a neurologist at the University of Chicago and editor of the *Journal of Comparative Neurology*, contributed "A Biological Survey of Integrative Levels" (1949). He referred to Novikoff's article for the notion of integrative levels and stated that synthetic methods must be used in science in addition to analytic ones: "This is because the new properties that emerge in directive activities are unique combinations integrated from elements which lack these properties." (1949: 226)

T. C. Schneirla (1902-1968), a comparative psychologist, also cited Novikoff in "Levels in the Psychological Capacities of Animals" (1949). Schneirla argued that the whole-part relationship was transformed with the appearance of higher levels. Units which became parts of wholes at higher levels had different qualities as subsystems of larger wholes than they had as wholes at their own level. Schneirla returned to this holistic aspect of his view of emergence in "The 'Levels' concept in the Study of Social Organization in Animals" (1951), where he stated: "A component functioning as a closely meshed unit in an organismic whole, or an individual organism functioning in a group, is qualitatively very different from the same component or individual when alone." (1972: 443)

Schneirla rejected the concept of societies as organisms, and considered the "supraorganism" concept of Emerson as reductionist, in agreement with Novikoff. Schneirla identified as "overlapping and interrelated adaptive stages" the physical, chemical, biological, biopsychological, psychological, subsocial, and social. The transition stages of the biopsychological and the subsocial were included to ensure that the psychological and social stages were clearly marked off from the preceding biological and the psychological ones, reinforcing the demarcation between distinct levels.

This brief history of the concept of integrative levels has highlighted the debate between organicists and anti-organicists, both of whom accepted emergence, at least from the inorganic to the organic, but disagreed on the relation between the social and the biological. Organicists tended to conflate the two, while anti-organicists like Novikoff stressed the novelty of the social relative to the biological.

(b) Renewal of Interest in Emergentism: from Meehls and Sellars to Peter Medawar

Emergence as a philosophical concept became a subject of renewed interest in philosophy of science as physicalist reductionism revealed its oversimplified project for the foundations of science. Unlike some philosophies which tend never to get beyond abstract preliminaries, logical positivism had committed itself to the concrete realization of the project it proposed, that of the reduction of all of science to physics. When it became evident that there were serious obstacles to the reduction even of chemistry to physics, let along of biology and psychology, the logical positivists were forced to conclude, using their own verificationist criterion for truth, that their project had failed. This was no more evident that in the second volume, issued in 1962, of the *Encyclopaedia of the Unified Sciences*, where the editors solicited an essay on philosophy of science from Thomas Kuhn. Initially intended as a just a chapter in the encyclopaedia, *The Structure of Scientific Revolutions* (1962) became the main text of post-positivist philosophy of science, which the editors of the encyclopaedia neither censored nor discouraged, and indeed, encouraged.[7] The effort of the logical positivists was an honest one, and this virtue led them to recognize the demise of their own system.

Of considerable significance for the return of emergentist philosophy to mainstream philosophy of science was "The Concept of Emergence" (1956) by P. E. Meehl and Wilfrid Sellars, the son of Roy Wood Sellars. This article appeared in the *Minnesota Series in the Philosophy of Science*, a widely read annual publication. Its aim was not to show that emergentism was true, but rather to show it to be a logically consistent concept which was worthy of consideration. The bulk of the paper was given over to the analysis of Pepper's 1922 argument – discussed earlier – to the effect that emergent laws were incoherent and emergent properties were epiphenomenol. Sellars and Meehl held that Pepper had overstated his original case concerning the two functions f_1 and f_2, where f_2 included variables representing emergent qualities and f_1 did not. Since the variables involved in both cases were q, r, s and t, Sellars and Meehl referred to processes in this "*qrst* four-space." Their claim was that f_2, the emergent law, held only for some areas of *qrst*-space, while f_1 held for all regions. The laws were different, at least in terms of their domains of application, and this disposed of Pepper's argument concerning the redundancy of emergent laws, though it did not yet settle the case in favour of the possibility of emergent qualities.

Sellars and Meehl noted that a survey of the emergentist literature indicated that sense data or secondary qualities, which they termed "raw feels', were the most likely candidates for emergent qualities. The last step of their paper was to introduce the distinction between physical$_1$ and physical$_2$. Any event or entity was physical$_1$ if it belonged in the space-time network, and was physical$_2$ if "it is definable in terms of theoretical primitives adequate to describe completely the actual states though not necessarily the potentialities of the universe before the appearance of life." (1956: 252) The emergentist claim could then be reformulated to state that raw feels were not physical$_2$ though they were physical$_1$: raw feels could not be completely defined in terms of concepts which applied to that stage of the universe before the advent of life, though they could be defined as functions in space-time of neuronal structures of the brain. This did not mean that raw feels were reduced to neuronal systems for being functions of them, since a functional relationship $y=f(x)$ does not necessarily equate y with x. On the other hand it did not exclude the possibility that raw feels might indeed be only neuronal states, in the case where f was the identity function. Sellars and Meehl concluded on this indecisive note: "Whether or not there are any emergents in the sense we have sought to clarify is an empirical question. Our only aim has been to show that Pepper's "formal" demonstration of the impossibility of non-epiphenomenal emergents is invalid." (1956: 252) Though they were themselves not proponents of emergentism they nonetheless considered it as a legitimate alternative, one worthy of consideration by other philosophers of science.

Herbert Feigl, in *The Mental and the Physical* (1958), adopted the Sellars-Meehl distinction between physical$_1$ and physical$_2$ and their terminology of "raw feels" for sensory qualities. Feigl's own preference was for mind-brain reductionism as contained in his "identity theory" where the relation between raw feels and neuronal systems was in fact that of identity. Nonetheless, though he believed that the psychological could be reduced to the physiological, he thought that this might not be the case for the organic relative to the inorganic, in opposition to Carnap's physicalism. In his contribution "Physicalism, Unity of Science and the Foundations of Psychology" to the 1963 *Library of Living Philosophers* volume on Rudolf Carnap, Feigl distinguished between two related but distinct aspects of Carnap's philosophy. The first was the thesis of the "unity of science" according to which there was both unity of the language of science and unity of the confirmation basis for all empirical claims of the natural and social sciences. The second was the thesis of "unitary science," a reductive claim according to which the "facts of the natural and the social sciences can all be derived – at least in principle – from the theoretical assumptions of physics." (1963: 227-228) Feigl considered the possibility of accepting the first of Carnap's theses, that of the unity of science, while rejecting the second, that of a unitary science: "Since emergence does not conflict with scientific determinism, or with the degree of determinism countenanced by modern physics, it is conceivable that the concepts and laws required for the explanation of biological and psychological processes will not be reducible to those sufficient for inorganic purposes." (1963: 265)

Carnap, however, disagreed, and defended an anti-emergentist position, arguing that emergence implied a breach in the continuity of nature which transgressed scientific principles. He rejected differences in kind and accepted only "differences of degree within a continuum" (1963: 884), a position which Darwin had earlier defended as well. Carnap held that it was arbitrary to draw a line of demarcation at one point in the continuous scale of increasing integration and claim that the mental and the conscious occurred thereafter but not before. At best he held that emergentism was a human recommendation for the maintenance of a linguistic dualism between physical and psychological terms.

Ernest Nagel, in *Structure of Science* (1961), distinguished between two claims of emergentism: that of novelty, according to which new properties occurred from time to time, and that of unpredictability, according to which these novel properties could not be predicted given knowledge of antecedent conditions and relevant laws. The novelty claim was one of an historical nature, and was distinct from the unpredictability claim, which was a logical question. Nagel concluded that emergence in the sense of the historical appearance of novel properties might have occurred, and that this could be accepted independently of the unpredictability claim. He also noted that the properties of some wholes – organisms – had not yet been reduced to the properties of their parts. This neither proved that they could never be so reduced, nor disproved that they might eventually be so reduced, and Nagel, like Feigl, W. Sellars, and Meehl concluded on this indecisive note.

Arthur Koestler and J. R. Smythies edited *Beyond Reductionism: New Perspectives in the Life Sciences* (1969) as the proceedings of a conference held at Alpach, Austria, whose aim was to criticize what Koestler perceived as a still dominant reductionism. His reaction was to develop a holistic and anti-reductionist position, centered on the concept of "holons," which, similar to Smut's wholes, drove the evolutionary process forward. A more balanced view was contained in the contribution by Paul Weiss, "The Living System: Determinism Stratified," where he argued that both the top-down and synthetic approach of holism, and the bottom-up and analytic approach of reductionism should be combined in what he termed "epistemological neutralism." This would recognize both the legitimacy of analysis into parts, and "the capacity of recombination of systems into what then appear as super-systems with the emerging properties of novelty and creativity." (1969: 33)

The volume edited by Francisco Ayala and Th. Dobzhansky, *Studies in the Philosophy of Biology* (1974) – the proceedings of a conference held in Bellagio, Italy – also contained a number of articles placing the accent on levels and emergence. Morton Beckner in "Reduction, Hierarchies and Organicism" (1974) considered that a phenomenon was emergent when the theory concerning it could not be reduced to theories about entities at levels below it, shifting the focus of emergence from real systems to theories about them. The same event could be described differently depending on the conceptual schema chosen. A theory which fully explained a process must not only take into account events at the level of that process, but also lower or higher level events which influence and affect it.

Peter Medawar, in his contribution "A Geometric Model of Reduction and Emergence" (1974), developed an idea he had earlier presented in *Induction and Intuition in Scientific Thought* (1969), where he had ranked geometric theories in an emergent hierarchy of metric geometry, plane geometry, affine geometry, and topology. There he noted that "(a) each geometry is a special case of its predecessor, i.e. is derived by imposing special restrictions upon or by defining a subgroup within the one preceding it; (b) all theorems of one geometry are also theorems in its successors; (c) new concepts (e.g. of parallelism, circularity or shape) "emerge" at each level which have no meaning and cannot be envisaged at an earlier level; and (d) there is a progressive enrichment in the number and variety of concepts and the particularity and degree of detail of the theorems." (1969: 19)

This was one of the few instances where an emergentist structure was postulated within mathematics itself, though Conger and Feibleman had earlier included the mathematical as one of the levels of emergent reality. Medawar extended his claim from the formal geometric series to the factual series of the sciences, and argued that the same kind of relationship he had identified for the hierarchy of types of geometry also held between physics, chemistry, biology, sociology, and ecology: "As we go down the line, the sciences become richer and richer in their empirical content and new concepts emerge at each level which simply do not appear in the preceding science." (1974: 61) Succeeding sciences were special cases of preceding ones in the following sense: only some molecular interactions formed the subject matter for biology, and only some biological interactions formed the subject matter for sociology. Laws of lower level sciences were still true at the higher level but often were no longer relevant – such as quantum mechanical effects at the social level – while higher level concepts did not occur in lower level sciences – such as economic exchanges at the chemical level.

(c) Supervenience and Downward Causation: Donald Campbell and Roger Sperry

A problem related to the emergence of levels of reality was the reverse action of the higher on the lower. This was known as the problem of "supervenience" in the terminology of Lloyd Morgan, and has been discussed in recent literature as "downward causation' as well.

Donald T. Campbell, a psychologist who has been active in developing theories of evolutionary epistemology, discussed this problem in his article "'Downward Causation' in Hierarchically Organized Biological Systems" (1974), also in the Ayala-Dobzhansky volume *Studies in the Philosophy of Biology*. Campbell agreed with the following two reductionist principles for hierarchically organized systems: higher level processes were constrained by and acted in conformity with lower level ones, and lower level processes were needed by higher level ones as means in order for them to achieve their goals. In this respect, Campbell differed from radical anti-reductionists or pure holists who argued that lower level phenomena were irrelevant to higher level emergent ones.

Campbell then added the two principles of emergence and downward causation which marked his disagreement with pure reductionism. He held that emergence operated at the organic level, which could not be reduced to the inorganic: "Biological evolution in its meandering exploration of segments of the universe encounters laws, operating as selective systems, which are not described by the laws of physics and inorganic chemistry, and which will not be described by the future substitutes for the present approximations of physics and organic chemistry." (1974: 180) Emergence set the stage for downward causation which was the return influence of higher level events, laws, or phenomena on lower level ones. Downward causation was a constraining factor, influencing the prevalence and distribution of lower level phenomena: "Where natural selection operates through life and death at a higher level of organization, the laws of the higher level selective system determine in part the distribution of lower level events and substances." (1974: 180) As so formulated, downward causation applied only to biological systems.

Campbell's emergentist concept of supervenience should not be confused with a very different use of the term by Jaegwon Kim in articles such as "Supervenience and Nomological Incommensurables" (1978) and ""Psychophysical Supervenience" (1982). Kim derived his notion of supervenience from the ethical theories of G. E. Moore and R. M. Hare, where a moral concept was said to "supervene" on physical ones. In Kim's theory, the supervening property, usually a mental one, varied along with the supervened property, usually a physical one. The supervened or lower level quality controlled the supervening or higher level one, which would be better described as epiphenomenal rather than emergent.

Roger Sperry, the neuroscientist who investigated split-brain patients, also used the term "downward causation', though in a more holistic sense than Campbell. In an early article, "Neurology and the Mind-Brain Problem" (1952), Sperry set the tone for his analysis when he remarked that mental phenomena were not found at the neuronal level, but were functions of higher level interactions of complexes of neurons. In "Problems Outstanding in the Evolution of Brain Function" (1964) he noted that evolution complicated the universe with "new phenomena, new properties, new forces"; older phenomena, properties, and laws were "superseded, overwhelmed, and outclassed," though they did not get "lost or cancelled." (1964: 2) In "A Modified Concept of Consciousness." (1969), though he argued that consciousness was a "direct emergent property of cerebral activity," an "integral component of the brain process," his emphasis was on the "top down control" or "directive holistic form of control" which the mind exercised over the brain. (1969: 532) He concluded: "In this holistic sense the present proposal may be said to place mind over matter, but not as any disembodied or supernatural agent." (1969: 533)

In stressing this holistic aspect, Sperry hoped to avoid what he considered as the excesses of materialism, which he associated with communism in politics and behaviourism in psychology. But holism had the effect of shifting his philosophy towards idealism, and in "Mind-Brain Interaction: Mentalism, Yes; Dualism, No" (1980), Sperry accepted the term "mentalist" to describe that aspect of his philosophy according to which mind, having emerged from brain, controlled it: "In calling myself a "mentalist', I hold subjective mental phenomena to be primary, causally potent realities as they are experienced subjectively, different from, more than and not reducible to their physico-chemical elements." (1980: 198) As a result of such formulations, particularly his use of the term "primary" in describing the relation of the mental to the physical, Sperry's position is better described as holistic and idealistic rather than as purely emergentist.[8]

The questions of supervenience and downward causation were only partially answered in the work of Campbell and Sperry, since Campbell limited himself to supervenience in biological systems, and Sperry shifted his emphasis to the mentalist or idealist position of "top-down control." A more adequate response to the question raised by supervenience is the following: whereas emergence is the appearance of qualitative novelty in synthetic wholes, and results in the movement from a lower to a higher level, supervenience, as the return action of entities and processes of a higher level on those of a lower one, is the quantitative modification of the distribution of lower level units. This concept of emergence is based on that of Campbell, but applies to all levels, not just the biological, and makes explicit the quantitative rather than qualitative nature of the process. Higher levels are not irrelevant to the lower levels, and so emergence is not epiphenomenal. Nor does emergence modify the laws of the lower levels, or assume a "top down control" as proposed by Sperry, since this shifts supervenience away from emergentism and skews it towards holism and idealism.

(d) The Autonomy of Biology: Ernst Mayr

Ernst Mayr, whose major contribution to biology was the inclusion of systematics in the modern synthesis, has used the concept of emergence in his argument for the autonomy of biology.[9] Mayr first used the term "emergence" in "The Emergence of Evolutionary Novelties" (1960). But despite its title, the article analyzed novelty in structural-functional terms of new properties and old systems: "The emergence of new structures is normally due to the acquisition of a new function by an existing structure." (1960: 377) In "Cause and Effect in Biology" (1961), Mayr expressed the principle of emergence more classically as follows: "When two entities are combined at a higher level of integration, not all the properties of the new entity are necessarily a logical or predictable consequence of the properties of the components."[10] (1961, in 1976: 369) This recognized that the emergence of new properties arose as the result of the integration of existing parts into new structures at a higher level, not as the result of old structures taking on new functions.

Though he adopted emergence in his philosophy of biology, Mayr was not opposed to reductionism in some of its forms. In *Growth of Biological Thought* (1982), he considered constitutive reductionism, explanatory reductionism, and theory reductionism as three distinct types. Constitutive reductionism claimed that the components of organisms were material and of the same nature as the components of inorganic compounds. Mayr accepted this form of reductionism and its view that the distinction between the organic and the inorganic was not a question of substance, but rather one of organization.

Explanatory reductionism claimed that a biological entity could not be understood unless analyzed into its basic component parts, macro-molecules in the case of organisms. Mayr recognized that this approach had been successful in a major case, that of the analysis by Crick and Watson of genes into their basis in DNA. But Mayr rejected the further claim that "all of biology is molecular biology," (1982: 60) because this neglected the fact that higher level systems were autonomous units, and as such were to some degree independent of lower level ones. Moreover, explanatory reductionism did not fully take into account the historical dimension of evolution, a second limitation to its application in biology.

Whereas he had given some credence to the first two types of reductionism, Mayr rejected the "theory reductionist" claim that biology could be shown to be a special case of physical and chemical theories. Though the description of a biological process involved the use of physical-chemical terms, it could only be understood by using specifically biological concepts as well. Predation could be described in the lower level terms of locomotion and capture, but could only be fully explained using further biological concepts such as digestion and survival of the fittest. Further, theory reduction neglected the existence of hierarchies in nature, and instead recognized only the physical level, failing to take into account the complexity and richness of reality.

Mayr distinguished two types of hierarchies in his philosophy of biology. The first was the constitutive hierarchy of macro-molecule, organelle, cell, tissue, organ, and organism, in which lower members of the hierarchy were physiologically combined to produce the higher members, which possessed emergent properties. The second type of hierarchy was the aggregational one, by which Mayr meant a taxonomic system such as that of species, genus, order, phylum, and kingdom. Unlike a constitutive hierarchy, an aggregational one was a strictly conventional classification where higher members of the series were not compounded from lower ones. The classification was conceptual, in the mind of the taxonomist, and therefore did not involve real emergent properties at higher levels. This analysis of aggregational hierarchies was in keeping with Darwin's own views on taxonomic categories as conceptual and not biological entities.

Emergentism was neither holistic nor vitalistic; rather, Mayr stated that "is a thoroughly materialistic philosophy." (1982: 64) Despite the controversy that once surrounded emergentism, Mayr's conclusion was: "So many authors, however, have now adopted the term "emergence" – and, like the term "selection" it has been "purified" by this frequent usage to such an extent (through the elimination of vitalistic and finalistic connotations) – that I see no reason for not adopting it." (1982: 64) He also adopted Campbell's term of "downward causation" for the return action of higher levels on lower ones.

Biology had become what Mayr termed an "autonomous" science through its emancipation from vitalist, mechanistic, and reductionist notions. Living systems differed from non-living ones in that they manifested complexity of structure linked with hierarchical organization and the emergence of new and unpredictable characteristics. In "Is Biology an Autonomous Science?" (1988) Mayr further distinguished the inorganic and the living worlds when he stated that evolutionary processes in the biological world were variational – characterized by random variation followed by natural selection – while those in the physical-chemical world were transformational – characterized by a sequence of states which arose causally one from the other. Moreover, he claimed that the biological sciences used probabilistic generalizations, rather than universal laws as in the physical sciences. In this he neglected stochastic process in quantum mechanics, some of which, as the source of mutations, are relevant to biology. In attempting to demarcate biology from physical science, Mayr tended in practice to neglect the importance of the physical-chemical levels as the basis for biological processes, even though, theoretically, he had allowed for constitutive reduction, and, to a lesser extent, explanatory reduction. Though Mayr's analysis of the biological level was emergentist, an integral emergentist theory should also take into account the material level, along with higher levels, including the social and the psychological.

(e) Factor Analysis of Emergence: Jonas Salk and Stanley Salthe

Just as Montague and others in the 1920s and 1930s attempted to analyze the concept of emergence and determine a mechanism underlying it, so have a number of authors in the recent period. A two factor or dyadic theory of emergence was advanced by Jonas Salk, one of the co-discoverers of the polio vaccine, in *Anatomy of Reality: Merging of Intuition and Reason* (1983).[11] Evolution occurred not only in the biological sphere, but also in the prebiological and the metabiological spheres. This triple distinction was similar to Spencer's inorganic, organic, and super-organic distinction, and Lloyd Morgan's three levels of matter, life, and mind. Salk stated that the mechanism of evolution was not the same in each sphere, as physical evolution involved an essential element of probability, as in quantum theories, biological evolution was due primarily to natural selection, while metabiological evolution was essentially characterized by conscious choice and the cultural transmission of ideas. He represented his theory of the three types of evolution in tabular form as follows: (1983: 32)

	Prebiological evolution	*Biological evolution*	*Metabiological evolution*
Emergence	Matter	Life	Consciousness
Unit	Atom	Cell	Mind
Components	Nucleus/Electrons	Gene/Soma	Intuition/Reason
Attributes	Interaction	Procreation	Creativity
Determinants	Probability	Selection	Choice

Table 13.1: Salk's Representation of the Types of Evolution

The cosmos itself emerged from non-manifest order to manifest order, leading to the next level, that of form characterized by the dichotomy between the continuous and the discontinuous. This led to the three spheres of factual reality, with each sphere subdivided into units or, in more traditional emergentist terminology, "levels." The physical sphere was divided into elementary particles and atoms, the biosphere into cells and organisms, and the metabiosphere into the human mind and culture. Each unit or level was characterized by binary components and was studied by a distinct science as follows: (1983: 31)

Unit	*Binary Components*	*Discipline*
Collective mind	Culture/Society	Sociometabiology
Mind	Intuition/Reason	Metabiology
Organism	Species/Individual	Sociobiology
Cell	Gene/Soma	Biology
Atom	Nucleus/Electrons	Chemistry
Particle	Energy/Mass	Physics
Form	Continuous/Discrete	Mathematics
Order	Non-manifest/Manifest	Metaphysics

Table 13.2: Salk's Levels of Reality

Each level featured the emergence of new qualities due to the interaction of polar opposites. This part of Salk's theory was reminiscent of Montague's attempt to analyze the two factors behind emergence at each level, and shared with it some of the same problems, that of oversimplification in particular. The emergence of elementary particles involves not only energy and mass, but charge as well. The intuition-reason distinction at the level of mind, though useful, neglected the role of sensation and perception. Nonetheless, the strength of Salk's treatment was his explicit recognition of levels within the three domains he identified.

Stanley Salthe developed a triadic theory of emergence in *Evolving Hierarchical Systems: Their Structure and Representation* (1985). Salthe, a biologist, collaborated with Niles Eldredge in "Hierarchy and Evolution" (1984), in an attempt to provide a philosophical basis for the theory of punctuated equilibrium which Eldredge and Stephen Jay Gould had earlier proposed in "Punctuated Equilibria: An Alternative to to Phyletic Gradualism" (1972). Salthe identified two major biological hierarchies: the ecological, which involved energy flow, and the genealogical, or reproductive hierarchy.[12] Salthe's interest in biological hierarchies led him to examine the concept of hierarchies in general, and he canvassed a number of theories of hierarchies going from elementary particles to communities of organisms at the biological level, and galactic systems at the cosmic level.

In any hierarchy of levels, he drew attention to its basic "triadic" structure: in order to represent the fundamental interactions of an entity, it had to be considered at its own level, analyzed into its parts at the next lower level, and considered in terms of the entities at the next higher level into which it entered via synthesis. The lower level provided constraints, which could be conceived of in a positive sense as possibilities or initiating conditions for the emergent event. The upper level also provided a constraint by acting as the environment and providing boundary conditions for the focal level event:

> What "emerges" we can take to be the product of some focal-level process(es). It therefore had to be among the possibilities engendered by permutations of possible initiating conditions established at the next lower level. ... What actually will emerge will be guided by combinations of boundary conditions imposed by the next higher level. (1985: 101)

The emergent was therefore sandwiched between lower and higher level constraints, and further investigation was required in any concrete case to determine their nature. This triadic theory was a potentially fruitful one, more so than Salk's dyadic concept, which was too specific and restrictive in its identification of binary components.

(f) Emergence and Three Worlds: Karl Popper

Karl Popper has had a major influence on contemporary philosophy of science through his theory that scientific progress proceeded through "bold conjecture" followed by "ruthless refutation," as set out in his *Conjectures and Refutations* (1963). Popper's first reference to emergence occurred in *Objective Knowledge: An Evolutionary Approach* (1972) when he situated his conjecture and refutation theory as an instance of a more general evolutionary process of trial and error.[13] Organisms had to solve the problem of their relation to the environment, and their efforts to do so were so many trial solutions, with natural selection acting as an "error-elimination" mechanism. The trials corresponded to conjectures, and the error-elimination mechanism to refutations. Popper's schema for such a process was the following: (1972: 243)

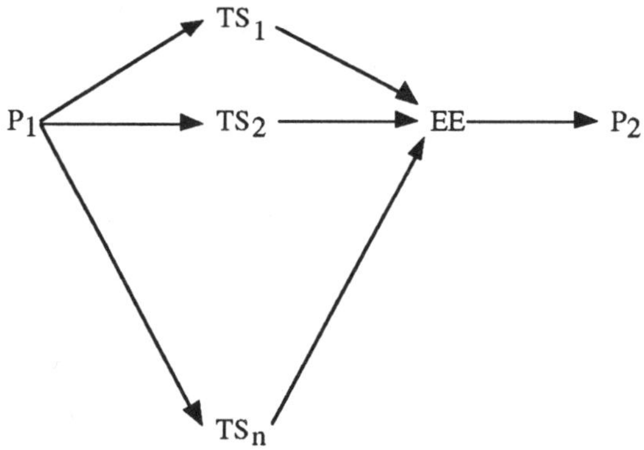

Diagram 13.1: Popper on Problem Solving

P_1 was the original problem to be solved, such as the biological problem of survival, or a theoretical problem in science; the TS_i were the various trial solutions, and EE was the error-elimination mechanism. The error-elimination mechanism did not produce a definitive solution to the problem, since this would amount to an absolute truth inconsistent with Popper's theory of fallible, tentative, and refutable knowledge. But it did winnow out the wrong solutions, and led, if not to a definitive solution, then to some new problem P_2, which emerged from the problem solving situation: "The theory hereby gives implicitly a rational account of what has usually been called by the somewhat dubious names of 'creative evolution' or 'emergent evolution'." (1963: 244) In "Scientific Reduction and the Essential Incompleteness of All Science" (1974), Popper situated his emergentist option more positively as follows: "We live, it appears, in a world of emergent evolution: of problems whose solutions, if they are solved, beget new and deeper problems. Thus we live in a world of emergent novelty; of novelty, which, as a rule, is not completely reducible to any of the preceding stages." (1974: 281)

In his "Natural Selection and the Emergence of Mind" (1978), Popper recognized four levels of emergence: the emergence of atomic nuclei and particles, the emergence of life, the emergence of conscious states, and the emergence of the products of mind. He also identified four stages in the emergence of consciousness: the warning state when pain or discomfort first appeared, a stage when imagined or vicarious trial and error replaced real trial and potentially fatal error, the stage of conscious aims, and the stage of language and the critical attitude towards one's own hypotheses.

However, at the same time that he adopted an emergentist ontogeny, Popper also argued for a pluralist ontology in his theory of three worlds. As developed in

The Self and Its Brain (1977), co-authored with the neuro-physiologist John Eccles, there existed three distinct but interacting substances: World1 of physical bodies and forces, World2 of minds, and World3 of the products of human minds, or culture. World3 was objective and its elements such as the integers existed independently of the human mind; propositions about these numbers were discovered, not invented. Mind at the intermediary level of World2 could interact not only with the brain of World1, but also with the culture of World3, and could do so directly, without any mediation by the brain. Such a claim for "brainless" mind-culture interactions was a peculiarity of the three world theory, and in contradiction to neuroscience, which recognizes corresponding brain states for every mental state.

But equally serious is the criticism that a pluralist ontology of the three world sort, assuming each world to be a distinct substance, was redundant once emergent evolution had been adopted. For having admitted the emergence of life from matter, and mind from living matter, it was superfluous to then postulate distinct substances of matter, mind, and culture as the three world theory did. Popper oscillated between interactionist pluralism and emergent evolution, and inconsistently combined them, as Lloyd Morgan earlier had attempted to combine neutral monism with emergent evolution. Emergent evolution requires an ontology which is consistent with its dynamic ontogeny, and this has been the task addressed by Mario Bunge.

(g) Emergent Materialism: Mario Bunge

Mario Bunge, who has contributed to the philosophy of the natural sciences, the social sciences, and psychology, and who has developed a systematic philosophy based upon the sciences, recognized emergence as one of the modes of change in *Causality: The Place of the Causal Principle in Modern Science* (1959).[14] However, from the outset he rejected one important aspect of previous emergent systems – the thesis of unpredictability – which he associated with irrationalism. In his most recent work, the eight volume *Treatise on Basic Philosophy* (1973-89) Bunge accepted what he termed "rational emergentism" as follows:

> The philosophy that combines an acknowledgement of emergence with the thesis that emergence is explainable and predictable within bounds may be called rational emergentism. This philosophy, foreshadowed by Sellars (1922), supersedes both atomism (though not its allegiance to

> science) and holism (though not its insistence on emergence), and it incorporates a critical realist theory of knowledge. (vol. 4, 1979: 251)

An important component of Bunge's ontology was the systems theory point of view.[15] A system was defined as the ordered triple $<C(x), S(x), E(x)>$, where $C(x)$ was the composition of the system, consisting of its component parts, $S(x)$ was the structure of the system, consisting of the relations entered into by any of the parts, with the internal relations being termed "connections," and $E(x)$ was the environment of the system, those systems of the external world with which the given system was related. Bunge's systemic ontology held that every real thing was a system or part of a system. Resultant and emergent properties were then defined as follows, with $p(x)$ the set of properties of x, and S the set of all systems:

> Let $P \in p(x)$ be a property of an entity $x \in S$. ... Then P is a resultant or hereditary property of x if and only if P is a property of some components $y \in C(x)$ of x other than x; otherwise P is an emergent or gestalt property of x. (vol. 3, 1977: 97)

The concept of a system provided a basis for distinguishing between holism, emergentism, and atomism. According to atomism, the properties of a system were just inherited properties of its parts. Holism asserted that the totality transcended its parts, and that the properties of the system were independent of those of its components. Bunge's emergentist view provided a *via media* between these two extremes. Some system properties were resultant while others were emergent, so that consideration of the properties of the parts was a necessary, but not a sufficient condition for understanding the system. In addition, the novel properties of the whole had to be examined.

While it has been held by holists that emergence and reduction were incompatible, Bunge in "Levels and Reduction" (1977) related them by distinguishing between their epistemological and their ontological aspects. The incompatibility held only between ontological emergence and ontological reduction, but not between ontological emergence and epistemological reduction. Epistemological reduction was a theoretical operation which did not alter the basic ontology: "In other words, reduction does not imply levelling: it relates levels instead of denying that they exist. Reduction, then, is a theoretical question that does not alter the level structure of the world." (1977: 79)

There were three possible views on reduction: anti-reductionism or holism, according to which knowledge of previous levels was not required to understand systems at a higher level; radical reductionism, or "atomism," the claim that facts and laws of any level could be fully reduced to those of the previous level; and moderate reductionism, "or the strategy consisting of reducing whatever can be reduced without however either ignoring emergence or persisting in reducing the irreducible." (1977: 80) Bunge opted for the intermediary position of moderate reductionism, rejecting the two extremes of holism and atomism. As a result, some theories were fully reducible to others, while some were only partially reducible. A theory T_2 was fully reducible to a theory T_1 if T_2 followed logically from T_1; in this case, T_2 did not refer to any emergent properties. T_2 was partially reducible to T_1 if it followed logically from T_1 and some non-empty set of auxiliary assumptions S concerning emergent properties.

In his theory of levels, Bunge argued that each of the metaphysical theses about the level structure of reality implied a corresponding epistemological thesis about knowledge of those levels of reality, each of which in turn implied a methodological principle concerning the procedures to use in order to understand levels.[16] The theses are summarized below, based on the exposition in *Method, Model and Matter*: (1973: 162-166)

Ontological theses	*Epistemological theses*	*Methodological theses*
1. Reality is a level structure and everything belongs to some level.	The level structure is knowable and science itself is a level structure.	Start be examining one level by itself and only then proceed to he others.
2. Some properties are gained and others lost in the course of emergence.	Every science has its peculiar objects as well as retaining some but not all of the objects of other sciences.	Try to explain emergence away, and failing to do so, take it seriously.
4. Every level has a degree of autonomy and stability.	Every level of science has a degree of autonomy and stability.	Begin with the examination of facts at one level and only then go to other levels.
5. Events are primarily determined by laws of their own and contiguous levels.	Events should be explained primarily in terms of their own and adjoining levels.	Start with intra-level laws, then proceed to inter-level laws.

Table 13.3: Bunge on the Metaphysics, Epistemology and Methodology of Levels

In *Scientific Research* (1967), Bunge recognized four basic levels: the physico-chemical, the biological, the psychological, and the socio-cultural. In

Treatise on Basic Philosophy (vol. 4, 1973), he modified his schema of the levels of reality. Though he argued that the mental was emergent relative to the physiological, and that psychology as a science was not fully reducible to physiology, he nonetheless did not acknowledge mind as a distinct level, but instead included it as a sub-level at the intersection of the social and the biological.[17] The four levels which Bunge identified were the physical, the chemical, the biological, and the social, with this latter branching from the biological along with the technological: (vol. 4, 1973: 250 and 46)

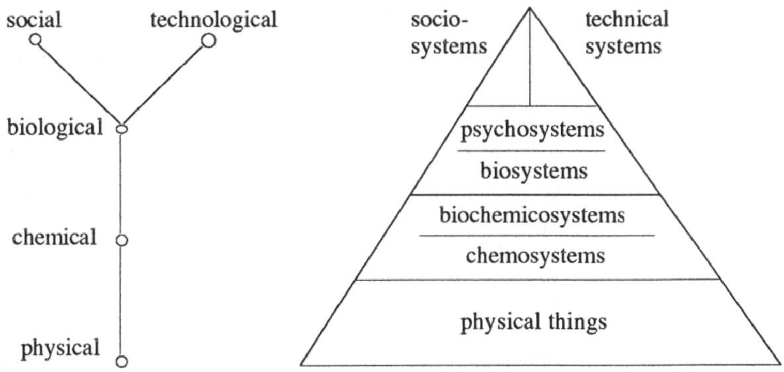

Diagram 13.2: Bunge's Recent Models of the Level Structure of Reality

Bunge's hierarchy of levels was based on the part-whole relationship: entities at a higher level were composed of parts from the immediately lower level. The biological and the social did not pose any difficulties for this analysis, since social systems were systems of organisms, and biological systems were systems of cells, whose parts were composed of macro-molecules. However, the mental level posed a problem, as minds were not systems of lower level entities, but rather were identified with special types of organismic sub-systems at the biological level. In "Emergence and Mind" (1977), the first article in a multi-author debate on the mind-body problem in the journal *Neuroscience*, Bunge stated his emergentist view on the mind-brain relationship as follows:[18]

> (i) All mental states, events and processes are states of, or events and processes in, the central nervous system of vertebrates; (ii) these states, events and processes are emergent relative to those of the cellular components of the CNS; (iii) the so-called psycho-physical relations are interactions between different subsystems of the central nervous system, or between them and other components of the organism." (1977a: 506)

There was a difference between Bunge's general definition of emergent properties, given on page 168, and his treatment of the emergent character of the mental in the above quote. In his general definition, Bunge analyzed emergents as properties of systems not possessed by any of their parts, but in the above quote he analyzed the mental in terms of "states, events, and processes." In so doing he considered the mental not as a property of certain sub-systems of the brain and central nervous system, but as those specific sub-systems themselves. This was not reductionist, as the neurons which made up these subsystems did not think; only systems of neurons could be considered as mental. But neither was it a direct application of the definition of emergence in terms of properties. The advantage of Bunge's modified approach was that he was able to analyze mind-body interaction as an interaction between two different types of biological subsystems, the one type designated as mental, the other not. The disadvantage was that it excluded the mental as a level of reality. The term "mental" could be applied only to certain dependent subsystems of the organism, while the molecular, the biological, and the social levels were composed of autonomous entities. The mental therefore did not have the same status as they did, and though not excluded from the emergentist ontology, it could not be considered as a distinct level of reality.

An analysis more consequent with the basic definition of emergence would be to consider the mental as a property of certain cerebral sub-systems, rather than as the sub-systems themselves, and analyze the levels of reality in terms of a property-based hierarchy, rather than a part-whole one. In this approach, to be defended in the conclusion, the mental level is composed of those organisms which possess the emergent properties of perception and conception. Entities making up that level are a subset of the set of organisms, consisting of those animals which have evolved brains.

Bunge has used a number of different terms to describe his ontological position: from "integrated pluralism" in his early writings, through "exact metaphysics" in the 1970s, to "scientific materialism" in more recent works. In *Scientific Materialism* (1982), he described his ontology as exact, as concepts were formulated in logical or mathematical terms; systematic, since all its hypotheses belonged to axiomatic theories; scientific, because its hypotheses were consistent with science; materialist, as every entity admitted by the ontology was material; dynamicist, since all entities underwent some change; systemist, because all things were systems or components of a system; emergentist, as every system possessed at least one property absent in its components; and evolutionist, since every emergence was the result of an evolutionary process.

The materialist aspect of the ontology was based on a dynamicist definition according to which an entity was a material object if and only if it could exist in at least two lawfully connected and distinct states: "An object is a material object (or entity) if, and only if, for every reference frame y, if $S_y(x)$ is a state space for x, then $S_y(x)$ contains at least two elements." (1982: 22) Otherwise, the object was immaterial, a non-entity. This definition did not presuppose any concept of stuff or substance, and excluded ghosts, immaterial minds, and entelechies as objects, since none of their supposed changes of states could be described in terms of laws of nature. Bunge's materialism reformulated an ontological position – everything is material, – in ontogenetic terms – anything is material which can undergo lawful change. In so doing, Bunge took up the challenge to develop a materialist basis for emergent evolution which Sellars and others had posed half a century previously.

CONCLUSION

The 19th century debate in England over the philosophical framework for evolutionary theory posed the question which Lloyd Morgan answered with his theory of emergent evolution, and his system served as the seed or core for the development of a philosophical trend. The debate among Darwin, Romanes, Wallace, Huxley, and Spencer concerned the continuous or discontinuous character of evolutionary change, the role of quantity and quality in that process, the relation between the organic and the inorganic, the relation between the natural and the supernatural, the mind-body problem, and the restricted or universal scope of the evolutionary process. This established the intellectual context for the work of Lloyd Morgan, who attempted to account for qualitative novelty within a continuous, natural, and universal evolutionary process. The result was the theory of emergent evolution.

The preceding analysis has identified the following theses as basic to this emergent evolutionary philosophy: firstly, evolution is a process of change which is productive of qualitative novelty in all domains of reality; secondly, qualitative novelty is the possession by a whole of emergent properties not possessed by any of its parts; and thirdly, reality can be analyzed as a structure of irreducible levels, each level consisting of wholes characterized by a new emergent property.

I have analyzed Lloyd Morgan's development of emergent evolution as a creative synthesis of a structural type – the result of his critical analysis, selection, and modification of the relevant aspects of the work, of the 19th century evolutionists, and the further influences of Mill, Lewes, Bergson, Spaulding, and Marvin. Lloyd Morgan incorporated the retained and modified theses in a theory which for the first time placed the emergence of qualitative novelty at the center of a naturalistic philosophy of continuous and universal evolution.

Nonetheless, there were difficulties with Lloyd Morgan's system. He attempted to combine his earlier neutral monist concept of the mental and the physical as correlated aspects of an underlying substance, with the newer concept of the mental as emergent from the organic. Moreover, he incorporated into his philosophy theses based not only on science, but on theology as well. I have argued that a more coherent and compact system of emergent evolution would have resulted had he limited himself to a science-based philosophy of emergence, and excluded the dual aspect component of his earlier neutral monism. The debate over emergentism dealt in part with problems in Lloyd Morgan's philosophy – its theological and teleological commitments in particular, – and in part with alternative views on the levels of reality, with further systems proposed by Sellars, Wheeler, Alexander, and Broad.

Emergent evolution during the period from the mid-1910s to the mid-1930s became a trend of thought, with both advocates and critics. Though temporarily eclipsed by reductionism during the period from the mid-1930s to the early 1950s, emergentism maintained its influence in the philosophy of biology through the related concept of integrative levels. With the decade of the 1950s, philosophers of science reconsidered emergentism as a legitimate alternative to reductionism and holism. In the last decades philosophers such as Karl Popper and Mario Bunge have incorporated emergentism in their philosophical systems, as have a number of biologists, including Ernst Mayr and Stanley Salthe, with results that I have traced in the last chapter. Emergent evolution was the outcome of a dynamic interaction of science and philosophy. The debate over emergent evolution is a continuing one, as philosophy of science, like science itself, cannot be content with fixed and rigid positions. In what follows, I provide my own conclusions concerning three issues raised in the course of tracing the history of emergentism: the part-whole relation, the analysis of emergence, and the levels of reality.

(a) The Part-Whole Relation

A number of different philosophies of the part-whole relation have been encountered, and Bunge (1977b) and Bahm (1981, 1983) have classified some of them. I propose the following analysis of the ontological and epistemological aspects of each of these philosophies – reductionism, mechanism, emergentism, organicism, and holism, situating them as a spectrum of views between the extremes of the first and the last.

Reductionists argue that the properties of wholes are just those of their parts, and that knowledge of the parts is both necessary and sufficient in order to understand the whole. Reductionism is either of the complete or restricted form. Complete reductionists, or physicalists, argue that all properties, whether chemical, biological, social, or psychological can be reduced to the physical, while restricted reductionists limit themselves to reducing properties of one kind to those of a simpler kind. Carnap's philosophy, which reduced all properties to the physical, was of the complete type, while Gerard's theory of societies as "epiorganisms" was of the restricted type, since it concerned only the reduction of social properties to biological ones.

Mechanists analyze the part-whole relationship in terms of cause and effect, arguing that the structure of the parts causes or brings about the properties of the whole. The kind or type of the effect is determined by the kind or type of the cause. Whereas reductionism operates at the item or "token" level of analysis – each property of the whole is to be found among those of one or more of the parts –

mechanism operates at the kind or "type" level of analysis – properties of the whole are of the same kind or type as those of the parts. Darwin adopted a mechanistic point of view in stating that brain caused the mind, just as the liver produced bile. Since the brain was material, so too was its effect – the mind.

Emergentists argue that there is at least one property of some wholes not possessed by any of their parts. Knowledge of the parts and their relations is a necessary but not a sufficient condition in order to understand the whole. Parts can exist independently of the whole, and novel properties of wholes can be lost via submergence when a system is reduced to its parts. Emergentism, as Bunge and others have pointed out, is not opposed to all forms of reductionism in science, when reductionism is understood as the attempt to analyze the structure of wholes and to explain their properties, as far as possible, in terms of those of their parts. When this fails, not merely in individual cases, but systematically, the unreduced properties are taken to be emergent. Philosophically, emergentists recognize qualitative novelty as a natural outcome of the evolutionary process, so that the world is more complex than the one level admitted by the physicalists.

Organicists recognize the existence of emergent properties of wholes, to which is added the claim that once a whole has appeared, its parts can no longer exist or be understood independently of that whole. Knowledge of the whole is a necessary condition for understanding the parts. When Needham affirmed that even atoms were to be understood on the organicist model, he did not claim that atoms were living beings, but rather argued that the parts of atoms – neutrons, protons, and electrons – could not be understood outside of atoms. Organicists also tended to interpret levels above that of life in terms of organisms, so that societies were seen as similar to living beings – "superorganisms" – a view defended by Wheeler and Gerard, but rejected by Novikoff and Schneirla.

Holists, like organicists, also accept the emergentist claim of novel properties of the whole, but go a step further and affirm that these properties can be understood without further consideration of the parts and their relationships. Knowledge of the parts is neither sufficient nor necessary. Smuts' holistic conception, similar to that of the gestaltists in psychology, took the basic unit of evolutionary change to be the whole. The holistic thesis is more radical than either organicism or emergentism, and occupies a position directly opposite to that of reductionism. Whereas for reductionism only the parts have to be analyzed, for holism only the whole has to be considered. Emergentism occupies a middle position between these two extremes with its thesis that knowledge of the parts is necessary but not sufficient for the comprehension of the whole. I summarize this spectrum of views on the part-whole relation in the following table (overleaf):

Philosophy	*Part/While: Ontology*	*Part/Whole: Epistemology*
Reductionism	Properties of wholes are always found among the properties of their parts – all properties are resultants.	Knowledge of the parts is both necessary and sufficient to understand the whole.
Mechanism	Properties of wholes are effects of the the parts and their structure.	Knowledge of the kind or type of the cause suffices to understand the type or kind of the effect.
Emergentism	Some properties of wholes are not the properties of any of their parts.	Knowledge of the parts is necessary but not sufficient to understand the whole.
Organicism	Parts cannot exist independently of a whole.	Knowledge of the whole is necessary in order to understand the parts and vice-versa.
Holism	The basic unit is that of the whole – wholes are independent of parts.	Knowledge of the parts is neither necessary nor sufficient to understand the whole.

Table 1: Theories of the Whole-Part Relation

(b) Qualitative Novelty

A symbolic analysis of emergence presents the concept in the clearest way possible, and following Reiser and Bunge, this requires the use of logical and set-theoretic terms. In order to analyze the concept of emergence, I represent a system a as the ordered triple of its components $C(a)$, properties $P(a)$, and relations $R(a)$. Let $S(a)$ represent the fact that a is a system. Then, $S(a) = \langle C(a), P(a), R(a)\rangle$, modifying Bunge's definition in that the set of properties is explicitly mentioned, while the environment is not. The component set of an entity or thing a is defined as the set of y's which are parts of a: $C(a) = \{y / y \text{ is a part of } a\}$. The analysis of a whole into parts depends on the results of science. The fact that electrons, protons, and neutrons are parts of atoms cannot be deduced logically from the concept of an atom, but must be discovered empirically by experiment. The thesis, now standard in particle physics, that the nucleons – neutrons and protons – are themselves made up of more elementary particles – quarks – is no less surprising than the fact that atoms are composite, and could not be deduced from the mere concept of "nucleon." Similarly, it was not always understood that the brain is composed of neurons, or that the cells of almost all multi-celled organisms consist of a nucleus and a cytoplasm. The part-whole relation requires empirical investigation in order to have specific content.

Let *Em(q,a)* symbolize the proposition that *q* is an emergent property of the system *a* and let *Sm(q,a)* symbolize the proposition that *q* is a submergent property of *a*. The basic thesis of emergence – that there exists a property possessed by a system which is not possessed by any of its parts – can be formulated as follows, following Bunge (1982):

$$Em(q, a) =_{def} [q \in P(a) \ \& \ (\forall y)(y \in C(a) \rightarrow q \notin P(y)].$$

A property *q* is submergent relative to *a* when it is a property of some part of *a* but not of *a* itself, as suggested by Baylis (1929):

$$Sm(q, a) =_{def} \exists y[y \in C(a) \ \& \ (q \in P(y)) \ \& \ q \notin P(a)].$$

A certain configuration of parts, or component configuration, is always associated with each emergent property. In the simplest case, this thesis states that an emergent property *q* of a system *a* is associated with some relation *r* between components y_1 and y_2 in *C(a)*. In a more complicated and realistic case, there would be associated with each emergent property *q* a set of relations r_1 through r_n holding between members of a subset of *C(a)*. A component configuration is invariably associated with every emergent property of a thing. Emergent properties are anchored in structures, and do not exist independently of them, though they are not reducible to them.

Every emergent property arises in a wider environmental context involving relations between the system and its environment. The environment *E(a)* of *a* is the set of external systems with which *a* interacts. Whenever an emergent property arises and persists, it does so in the context of interactions between the system and parts of its environment. For example, a cell must take in nutrients from its environment or else reproduction is impossible. The theses of component configuration and environmental context supplement that of qualitative novelty to form the core of emergentist claims about novel properties.

I have not considered the claim of unpredictability as essential to emergentism. The two concepts are separable in principle, for the following reason: emergentism is an ontological claim about the relation between the properties of wholes and the properties of their parts in the context of the evolutionary process, while prediction is an epistemological thesis about the ability of humans, given information about those parts, to predict the properties of the wholes at a given moment in the development of science. These two claims belong to distinct domains, and there is no *a priori* reason why one should necessarily imply the other. It may be the case that the properties of a chemical compound such as water cannot be determined given complete knowledge of the properties of its component atoms, as was claimed by many emergentists who considered the example. But the irreducibility of the problem may be due to the difficulty of the quantum mechanical computations involved, in which case the problem would be reducible in theory, even if in practice it could not be solved using exact methods.

Similarly, the development of biochemistry may enable scientists to predict under what conditions life emerges from its chemical precursors, but this would not change the situation that cells and organisms have properties not possessed by their macro-molecular constituents. Finally, even if some emergent properties are not predictable, this does not mean that emergentism denies the rationality of science. Recent developments in "chaos" theory have given simple examples of lawful non-emergent events which are unpredictable, such as complex weather patterns where slight differences in initial conditions lead to bifurcations and very different end-results. In this case, a deterministic process is unpredictable, yet undoubtedly an object of science.

An unresolved problem was that of the source of emergence. Numerous proposals have been made by the various philosophers examined in the course of this book: "*nisus* towards deity" and "natural piety" by Alexander and Lloyd Morgan, "complexity of organization" by Sellars, though he also took refuge in the "natural piety" of Alexander and Lloyd Morgan, "epitomization" by Conger, "sociality of the present" by Mead, "dual factors" by Montague, "statistical fluctuations" by Reiser, "local recapitulation of global levels" by Boodin, "formation of wholes" by Smuts, "dyadic interactions" by Salk, and "triadic structure" by Salthe. A complexity based theory of emergence, such as that alluded to by Sellars but not developed further by him, holds that at specific degrees of complexity of organization, new properties are formed in order to establish a fresh and simpler point of departure. Such a process in nature maintains a balance between growing quantitative complexity and emerging qualitative novelty. The challenge is to develop a common measure of complexity over the inorganic, organic, social, and mental domains. The difficulties involved in such an interdisciplinary project are only now being addressed. At the same time, Salthe's triadic theory is promising, focusing attention not only on the emergent and its lower level basis, but on its upper level boundary conditions as well.

The converse of the emergence of higher levels from lower levels is the supervenience of the former upon the latter. Lloyd Morgan analyzed supervenience as the return action of emergent processes upon their underlying components which introduced a "touch of novelty" to them, but did not examine whether that novelty was quantitative or qualitative. Campbell stated that supervenience was the effect of an emergent level on the prevalence and distribution of the lower level component parts, though he restricted this to life and death biological processes. This should be distinguished from the stronger holistic claim that the emergence of a higher level changes the laws of the lower level, or exercises top-down control in the sense of Sperry. Emergentism recognizes the autonomy of lower level processes, which can exist in the absence of higher level ones, and whose laws of operation are independent of those of the higher level. Emergence, the appearance of new properties in systems which integrate lower level components, is qualitative, while supervenience, the return action of these emergent structures on lower level entities, is quantitative.

(c) The Levels of Reality

Emergent evolution results in the establishment over time of a structure of distinct and irreducible levels of reality. Lloyd Morgan set the tone for later debate by including matter, life, and mind as the essential emergent levels, while others, such as Sellars, Wheeler, and Bunge, have included the social as well. In what follows, I present some concluding remarks on the structure of these four levels, including their major sub-levels.

The material or physical-chemical level is the base level and consists of sub-sub-atomic particles, atoms, molecules, and macro-molecules. Atoms occupy a central position as they can be analyzed into their sub-atomic components and synthesized into molecules. The sub-atomic particles are themselves quanta of the basic fields, presently identified – and partially unified – as the electromagnetic, the weak nuclear, the strong nuclear, and the gravitational. Molecules, which are stable compounds of atoms, are further combined to form macro-molecules.

The biological or cellular-organismic level arose next. Cells are made up of cell-components which are themselves made of macro-molecules, providing a link between the biological and the material levels. At the biological level, the properties of nutrition, growth, and reproduction have emerged. Though living units below that of the cell exist, in particular viruses, the most elementary unit capable of independent self-reproduction is the prokaryotic cell. Lynn Margulis in *Origin of Eukaryotic Cells* (1970) argued that the eukaryotic cell, which has its DNA encapsulated in its nucleus, evolved from symbiotic associations of prokaryotic cells. Eukaryotic cells are the basic components of multi-celled organisms – fungi, plants, and animals, including humans.

The social or population-ecological level is the third level. The simplest community of organisms is a population: interacting organisms of the same species which form a reproductive community. Ecosystems are systems of interacting populations of different species, and the system of all ecosystems is the biosphere. The link between a population and the preceding biological level is that each organism making up a population is composed of cells.

A population whose members can be differentiated into castes or classes according to distinct functions in the collectivity is a society. In the case of the social insects – ants, termites, and some species of wasps and bees – these societies are fixed, and the members of different castes have distinct morphological characteristics. In the case of the higher primates there is greater variability in societal structure, and in the case of humans, the greatest variability. The recognition of the biological basis of the social does not imply that social evolution is uniquely determined by the biological level, though it does impose constraints on the possibilities of social organization, since at the very least societies must meet the biological needs of the individuals which compose them.

Within some of the socially organized organisms – the higher primates and humans in particular – the further level of the mental or perceptual-conceptual has arisen. The precursor to mind is sensation, which appears in animals which possess brains, rather than in plants and fungi, which do not. Harry Jerison in *Evolution of the Brain* (1973) introduced the concept of "encephalization" to describe a major evolutionary shift at the level of the brains of birds and mammals: the development of cerebral capacity not directly tied to immediate bodily functions. He argued that this excess brain capacity was used to coordinate and fuse sensations into perceptions. The mental space of perception provides a model of the environment, an internal representation of external reality. As indicated by Popper, the evolutionary advantage is that an animal can consider choices in the model as to appropriate actions to take in reality, rather than be limited to real trial and potentially fatal error. The ability to choose among alternative acts is a manifestation of intelligence, and the awareness of these choices is consciousness, as suggested by Romanes.

Among the higher vertebrates perception passes into cognition, where intelligence is supplemented by the ability to infer the results of choices. The future, even the distant future for humans, is represented in the internal model of reality. Such capacities are found in the higher primates and humans, and perhaps exist as well in the cetaceans – whales, porpoises, and dolphins – all of which have brains with developed neo-cortexes. Finally, among humans, there is the lateralization of cerebral function which is characteristic of the development of language. The self, which can be given a name, becomes an integral, often central element of the internal representation of reality. Such a mind has the ability to construct in imagination alternative worlds, and is expressed by humans in cultural forms such as art and literature. Humans, through the use of their mental faculties, can investigate the nature of reality through science, and produce technologies in order to amplify bodily and cerebral functions. Minds are not disembodied entities, since "mind" in this sense refers to the set of emergent properties of the cerebral cortex of some socially organized organisms - the higher primates and humans in particular - assuring a link between the mental level and the preceding social and biological ones.

No entities other than material ones enter into the constitution of systems which are members of higher levels. This is the materialist ontology underlying emergent evolution, and is a reductive aspect of the philosophy. Any system at a level above the material can be decomposed into, or reduced to, components from the material level, ultimately atoms and their further sub-atomic constituents. However, the set of properties of entities at a higher level is not reducible to those of a lower level, but contains at least one new, emergent property. The advantage of emergent evolution is therefore double: on the one hand, it is monistic in affirming the existence of only one type of system component, while on the other hand, it is pluralistic in recognizing the existence of many types of novel properties. These new properties are qualitative novelties that have arisen in the course of evolution, and as a result of the appearance of significant biological, social, and psychological ones, reality can be analyzed as a structure of irreducible levels: matter, life, society, and mind.

NOTES

Chapter One

[1]Of the works published during his lifetime, four constitute Darwin's main contribution to evolution theory. These are *On the Origin of Species by Means of Natural Selection, or the Preservation of the Favoured Races in the Struggle for Life* (1859), *The Variation of Animals and Plants Under Domestication* (1868), *The Descent of Man, and Selection in Relation to Sex* (1871) and *The Expression of Emotions in Man and Animals* (1872). After initial work in the late 1830s and the early 1840s, Darwin returned to work on his theory of evolution in 1856, and had produced over a thousand pages of text when the letter from A. R. Wallace arrived presenting Wallace's very similar theory of evolution. After the joint presentation to the Linnean Society of London of excerpts from Darwin's 1842-44 work and Wallace's articles and letters of the 1850s, Darwin then produced an "abstract" of his own work, which became the *Origin of Species.* The much longer work that he was producing at the time has subsequently been published in several parts. Two chapters of Darwin's 1856-58 manuscript were published during Darwin's lifetime as *The Variation of Animals and Plants under Domestication*, while the rest was published by R. C. Stauffer under the title *Charles Darwin's Natural Selection* (1975). This work is of interest because it contains many footnotes by Darwin referencing the source of his ideas and facts. These, along with many digressions, were not included in the shorter *Origin of Species*, in keeping with its role as an "abstract."

[2]Darwin accepted that characters acquired during the life of an individual could be transmitted to its progeny. Unlike Lamarck, however, he rejected the notion that in the higher animals this was mediated by an act of willing. Rather, he sought a physiological mechanism of inheritance capable of producing such a result, and the theory of pangenesis was the result. According to this theory, as set out in *The Variation of Plants and Animals under Domestication* (1868), tiny particles, or "gemmules" are produced by each bodily organ and encapsulate its state of development. These gemmules congregated in the sexual organs, to be passed on through procreation and activated in progeny. Activation resulted in the parental character being reproduced in the children, providing the basis for a belief in the acquisition of acquired characteristics. Darwin was concerned about the speculative nature of his thesis, and referred to it in letters as a "working hypothesis." He maintained it because it allowed him to include a large number of facts of heredity under a general hypothesis. The subsequent demonstration by August Weismann

and others that the theory of pangenesis was false provides an example of a part of Darwin's theory that has been refuted.

[3]Darwin provided lists of factors of variation in his *Essays* of 1842 and 1844, *Natural Selection*, *Origin of Species*, and *Variations of Plants and Animals*. In the chapter "Laws of Variation" in *Natural Selection*, his list was as follows: the immediate or direct action of external conditions, acclimatization, effects of use and disuse of structure, correlation of growth, and compensation or balancing. In the *Origin of Species*, a similar, but not quite identical list was given at the head of the chapter on laws of variation. Darwin also considered that wide-ranging, much diffused, and common species varied most, as did species in larger genera as compared to those in smaller ones.

[4]In *An Essay on the Principle of Population* (1803), Malthus argued that human population increased more rapidly than food supply, resulting in over-population, and the consequent lowering of the living standard. Malthus took this as an argument against revolution, the French in particular, as there was no way to permanently improve the living conditions of the mass of the population. Whereas Malthus restricted his work to human populations and was concerned mainly with political implications, Darwin extended the concept to all species, and drew the biological conclusion that the ensuing struggle for existence was the condition upon which natural selection acted.

[5]In a hypothetical and simplified example, the characteristic might be the length of the claws of an animal that procured its food by digging it up from the ground. If the claw length were too small, the animal would not be successful in feeding itself, while those animals with longer claws would succeed and "prosper," as measured by the number of their progeny.

[6]In *Natural Selection*, Darwin was explicit in indicating that Milne-Edwards' concept of biological division of labor suggested his notion of divergence of character: "The doctrine is in fact that of the 'division of labour,' so admirably propounded by Milne Edwards, who argues that a stomach will digest better, if it does not, as in many of the lowest animals, serve at the same time as a respiratory organ; that a stomach will get more nutriment out of vegetable or animal matter, if adapted to digest either separately instead of both." (1856-58: 223) Nonetheless Darwin rejected Milne-Edwards' opposition to evolution, and he noted that the same concept, that of division of labor, could be used in both an evolutionary and non-evolutionary context. Biological division of labor, which Darwin also referred to as the "law of economy," reappeared in modified guise as divergence of character in Darwin's mature theory, adding an element that was absent from his 1842 and 1844 *Essays*, though present in Wallace's 1855 paper. At least one commentator (Brackman, 1980) argued that Darwin took the notion of divergence of character

from Wallace without acknowledgement, and that Wallace, rather than Darwin, was the "true" inventor of the modern theory of evolution. But if Darwin took the idea of divergence of character from Wallace, then he also took it from Milne-Edwards, and on this point the acknowledgment was explicit in *Natural Selection*.

[7]Darwin stated in his text accompanying the diagram of the tree of evolution: "But I must here remark that I do not suppose the process ever goes on so regularly as is represented in the diagram, though in itself made somewhat irregular, nor that it goes on continuously; it is far more probable that each form remains for long periods unaltered, and then again undergoes modification." (1859, 1: 168) The term "continuous" referred to the rate of change, rather than the process of change, and Darwin admitted that periods of stasis might be followed by periods of rapid change – as Eldredge and Gould (1972) would generalize in their theory of punctuated equilibrium. At no time did Darwin recognize non-continuity in the process of change, since this would amount to saltationist leaps which he held to be inimical to his theory.

[8]Darwin's main writings on plants were: *On the Various Contrivances by which British and Foreign Orchids are Fertilised by Insects* (1862), *On The Movements and Habits of Climbing Plants* (1865), *Insectivorous Plants* (1875), *The Effects of Cross and Self-Fertilization in the Vegetable Kingdom* (1876), *The Different Forms of Flowers on Plants of the Same Species* (1877), and *The Power of Movement in Plants* (1880).

[9]The term "biopsychism" was used by Ernst Haeckel in "Our Monism: The Principles of a Consistent, Unitary World-View" (1892), where he distinguished between panpsychism, biopsychism, and zoopsychism as three theories of the scope of the mental – ranging from panpsychism as universality of mind to zoopsychism which restricted mind to animals alone.

[10]Aristotle's theory analyzed terrestrial matter in terms of the four basic elements - earth, air, water, and fire – while a fifth element - aether - was reserved for celestial phenomena. Each of the four earthly elements was based on two non-contradictory qualities - hot or cold, wet or dry. For example, water was wet and cold, and could be transformed into air by replacing the quality of cold with that of hot, producing the element which was both wet and hot - air. The quantity of "hot" or "cold" required was irrelevant, since all that counted was the fact of the qualitative replacement.

[11]Darwin's quantitative view of the character of change was related, though not the same, as his defense of the thesis of gradualism. Gradualists hold that species form slowly over very long periods of time. This is an empirical consideration, and can be confirmed or refuted by measuring rates of speciation.

Rapid speciation followed by long periods of stasis can occur within a continuous evolutionary process, as in the punctuated equilibrium of Eldrege and Gould (1972). The condition is that rapid speciation occur over a finite, albeit small period of time, and not occur in an "instant" or just one generation, as hypothesized by the saltationism of Goldschmidt's (1940) "hopeful monster." It is possible to have a continuous but non-gradualist theory of evolutionary change, and the Gould-Eldredge punctuated equilibrium theory need not be considered as saltationist.

[12]The problem of the status of "species" has been the subject of considerable debate, an example of which is Michael T. Ghiselin, "Species Concepts, Individuality, and Objectivity" (1987), as well as the reply by Ernst Mayr, "The Ontological Status of Species: Scientific Progress and Philosophical Terminology" (1987), both in *Biology and Philosophy.* The problem of the status of species has also been the subject of a long lasting debate in the pages of *Journal of Systematic Biology.*

[13]A simplified example of Darwin's quantity-quality distinction can be illustrated as follows: consider two populations of animals co-descended from a single parental stock, which differ in that members of population *X* are on the average 2" taller than members of population *Y*. Suppose that one biologist grouped the two populations as sub-species of the same species, while another grouped the same two populations as distinct species within the same genus. In both cases the two biologists measure a height difference of 2" between the average members of the two populations. This quantitative measurement is not in dispute. But the first scientist has set 1.5" as the dividing point between species, while the second has set 2.5". As a result, they group the populations differently. The first biologist would call both populations by the same species name, whereas the other would not, recognizing a common name only at the level of the genus. This amounts to a qualitative distinction between the two taxonomic systems, even though neither the fact of common descent, nor the degree of quantitative modification of the two populations are in dispute. However, given the common descent, it would be impermissible to conclude that the two populations were not directly related and place them in separate kingdoms.

[14]In physical science the Boyle lecture series on natural theology continued throughout the 18th Century to bolster this interventionist view of God. No less a scientist than Newton actively subscribed to this program. In his *Mathematical Principles of Natural Philosophy* (1729), he criticized Descartes' vision of a God who, having created the universe, retired from active intervention. In his letters to Richard Bentley, the first Boyle lecturer, Newton stated that God might continue to

intervene in order to make sure that planets or stars did not collide as the result of gravitational attraction. (Cohen, 1975)

[15]The Rev. Francis Henry, Earl of Bridgewater, died in February of 1829 and left the sum of 8000£ sterling to be divided among eight authors for the purpose of publishing a series of volumes "On the Power, Wisdom and Goodness of God, as manifested in the Creation," to be illustrated by arguments from all branches of natural philosophy. The following volumes were published in this series: Rev. Thomas Chalmers: *On the Power, Wisdom and Goodness of God as Manifested in the Adaptation of External Nature to the Moral and Intellectual Constitution of Man*, John Kidd, MD: *On the Adaptation of External Nature to the Physical Constitution of Man*, Rev. William Whewell: *Astronomy and General Physics Considered with Reference to Natural Theology,* Sir Charles Bell: *The Hand: Its Mechanism and Vital Endowments as Evincing Design*, Peter Mark Roget, MD: *On Animal and Vegetable Physiology*, Rev. William Buckland: *On Geology and Mineralogy*, Rev. William Kirby: *On the History, Habits and Instincts of Animals*, and William Prout, MD: *Chemistry, Meteorology, and the Function of Digestion, Considered with Reference to Natural Theology.* Charles Babbage produced a ninth and unauthorized Bridgewater Treatise.

[16]The progress and development of Darwin's thought has been the subject of much study, especially since the publication of his *Notebooks* in the mid-1960s. There are four notebooks on evolution, Part I (July 1837 - February 1838), Part II (February to July 1838), Part III (July 15, 1838 - October 2, 1838), Part IV (October 1838 to July 10, 1839). Gavin De Beer also edited Part V - Addenda and Corrigenda, and Part VI - Pages Excised by Darwin. The notebooks were published in the *Bulletin of the British Museum (Natural History),* during 1960-67.These notebooks, produced during the period from 1837-41, reveal not only the germs of an evolutionary theory, but also provide insights into the philosophical influences on Darwin. The positions he held at that time were carried forward to his later writings, and helped shape his outlook on reality.

[17]Barrett indicated crossed out sections by placing them between "/" markers. Barrett has shown that Darwin read a review of Comte's *Positive Philosophy* in the *Edinburgh Review* for July 1838. The reviewer quoted textually from Comte on the three stages of the evolution of thought on page 280 of that article.

[18]Samuel Clarke in 1707 had argued against a materialist view of the mind-brain relation in his debate with Anthony Collins. Collins, in defending his position, held that the brain as a whole had the property of thought which none of its parts possessed. The debate was discussed in John Yolton, *Thinking Matter: Materialism in Eighteenth Century England* (1983), pp. 39-44.

[19]Priestley also argued for a theological basis for his materialism. The Bible spoke of the resurrection of the body, and not the soul, on the day of judgment. How would the mind and so the whole person be restored? This could take place only if the mind were simply a product of bodily organization, that of the brain in particular. The reconstitution of the body would be, in the same act, a restoration of that body's mind. Thus did a materialist ontology aid, not hinder religion. Needless to say, Priestley, both in his proto-emergentist and theological arguments, was an exception in England.

[20]Lamarck, in his *Zoological Philosophy*, developed a materialist theory of the mind-brain relation. Mind, as a capacity for thought, was the product of cerebral organization. In an animal with only afferent-efferent connections and no central brain mass, movement alone was possible. Sensibility was achieved by animals whose nervous connections had consolidated in a single cerebral mass. Finally, thought, as the highest embodiment of mind, was characteristic of brains with a granulated neo-cortex, as in humans.

[21]Stephen Jay Gould in "Darwin's Delay" (in 1977a) suggested that Darwin's materialism was the reason for his delay of some two decades, from the mid-1830s to 1859, in publishing his theory of evolution. Though certainly a factor, this claim appears exaggerated if taken as the sole reason for the delay, for the following reasons. Firstly, Darwin's theory was not yet fully developed in 1838 despite his having conceived of the concept of natural selection. In particular, the problem of divergence of character was not worked out until the mid-1850s, following the suggestions by Milne-Edwards, and Wallace as well. Part of the delay, then, was in order to flesh out the theory beyond the mere concept of natural selection. Secondly, Darwin typically published his novel results only when at least one other scientist had come to the same conclusion. His first statement of evolution by natural selection was prompted by Wallace's arriving independently at the same conclusion and being ready to publish; Darwin's work on human evolution in *Descent of Man* was published after Lyell, Huxley, and Haeckel had already published evidence for substantially the same conclusions; and *Darwin's Expression of Emotions in Man and Animals* belatedly treated of comparative psychology once he had Romanes as a collaborator. Darwin's delay at the personal level might have been based on the desire "not to go it alone" as he waited on at least three major occasions for supporters ready to publish along the same lines as his own theories.

[22]"Man" was a gender-specific concept in the work of Darwin and the other evolutionists of this period, with one sex taken as the representative of the whole species. The ideological component is that of the superiority of males in comparison with females, a cultural bias widespread in the 19th century. In 20th

century terms, it is more appropriate to speak of "humanity", understanding as we now do that gender hierarchy is a social construct and not a biological necessity. In discussing Darwin's presentation of his own theory, the male-specific references are maintained for the sake of historical accuracy.

Chapter Two

1Spencer's major work, the 5 part, 10 volume *Synthetic Philosophy* (1893-1902) was published in installments from 1855 through to 1893, a period of some 38 years. It consisted of the following parts: *First Principles* (1867), *Principles of Biology*, 2 vols. (1864, 1867), *Principles of Psychology*, 2 vols. (1870, 1872), *Principles of Sociology*, 3 vols. (1876-1896), and *Principles of Ethics*, 2 vols. (1879-1893). As the titles indicate, Spencer aimed to encompass the basic domains of knowledge, with a general introduction setting out the evolutionary and philosophical principles of his work – *First Principles* – followed by systematic treatment of the basic principles of biology, psychology, sociology, and ethics. The project and ordering of the sections is not without recalling Auguste Comte's *Positive Philosophy*, and this problem of the relation between Spencer's theories and those of his French predecessor was to be a vexed one for Spencer, leading to polemics with the positivists over priority and originality.

2In the United States, Spencer's disciple E. A. Youmans was editor of the *Popular Science Monthly*, which regularly published articles on the debate over evolution in Europe and America. Youmans also edited for the publisher D. Appleton and Co. a 60 volume collection of reprints of the works of Darwin, Spencer, Huxley, and Tyndall, as well as miscellaneous volumes on evolution by Haeckel and others. Spencer's work was as a result widely known in America as well, and situated as one of the canonical texts both in evolution theory and systematic philosophy.

3In *Origin of Species* Darwin stated: "I have called this principle, by which each slight variation, if useful, is preserved, by the term Natural Selection, in order to mark its relation to man's power of selection. But the expression often used by Mr. Herbert Spencer of the Survival of the Fittest is more accurate, and is sometimes equally convenient." (1859: 99-100) In the model of Darwin's theory in the preceding chapter, I have reserved the term "natural selection" for the causal factor which, when acting upon a population in a state of competition for scarce resources, resulted in the survival of the fittest. In this sense, the term "survival of the fittest" referred to the effect. This, I believe, is consistent with the spirit of Darwin's *Origin of Species*, which was based on a mechanistic theory of cause and effect relationships.

[4]Spencer further expanded his idea of evolution in two subsequent articles: "Progress: Its Laws and Cause" (1857) and "Transcendental Physiology" (1857). Its exposition also formed the subject matter of chapters XII - XXII of *First Principles*, and articles such as "The Factors of Organic Evolution" (1886), where he compared and contrasted his concept of evolution with those of Darwin and Lamarck.

[5]Robert Chambers, an Edinburgh book and journal editor defended a neo-Lamarckian view of biological evolution, and extended this evolutionary concept to a general theory of change throughout all nature. He saw a conformity between the Kant-Laplace nebular hypothesis and Lamarck's theory, since both held that phenomena - whether astronomical or biological - progressed or developed through stages. The fact that Chambers published the work anonymously added to its aura of forbidden fruit, and even T. H. Huxley denounced it at the time, for confusing natural and supernatural processes in the concept of "creation by law."

[6]Spencer's ten volume *Synthetic Philosophy* was a daunting work for the neophyte to read. A disciple, F. Howard Collins, wrote *An Epitome of the Synthetic Philosophy* in 1889, which was published by Spencer's American publisher, D. Appleton and Co. Spencer himself contributed a resume of his own philosophy of evolution in 16 points, which was printed as the preface to Collins' work.

[7]Although few readers made it through the full ten volumes of the *Synthetic Philosophy*, many more would be familiar with this volume, *Data of Ethics*, as well as the introductory *First Principles*, which were the most widely available parts of the system. The turn of the century American editors of "A Library of Universal Literature" included both these works among the first dozen volumes of the series on science. Also included were Darwin's *Origin of Species, Descent of Man*, and *Journal of Researches,* Huxley's *Science and Education*, and Tyndall's two volumes of *Fragments* on physics. The series was published in New York by Collier and Sons and circulated widely.

Chapter Three

[1]Replying to a letter from Darwin, Huxley explained why he had not mentioned the term natural selection in an essay of 1880: "I hope you do not imagine because I had nothing to say about "Natural Selection," that I am at all weak of faith on that article. On the contrary, I live in hope that as paleontologists work more and more ... we shall arrive at a crushing accumulation of evidence in that direction also. But the first thing seems to be me drive the fact of evolution into people's heads; when that is once safe, the rest will come easily." (1903: ii:

279) Given that the recipient of the letter was the author of the theory of natural selection, Huxley's words were diplomatic. His reference to "weak of faith" from a man weak in all faiths was perhaps rhetorical, though his use of the term "fact of evolution" indicated his stronger belief in the objective existence of the process as a whole.

[2]Huxley was not alone in speculating about saltationism in evolution. Charles Lyell, author of the *Principles of Geology* (1830-33) also had his doubts. His *Geological Evidences of the Antiquity of Man* (1861) was one of the first book-length treatments of the evolution of *homo sapiens*. In the conclusion to the first and second editions, he used the term "leaps" to describe the transition from lower to higher intelligence in the evolution from animals to humans: "If in conformity with the theory of progression, we believe mankind to have risen slowly from a rude and humble starting-point, such leaps may have successively introduced not only brighter and higher forms of grades of intelligence, but at a much remoter period may have cleared at one bound the space which separated the highest stage of unprogressive intelligence of the inferior animals from the first and lowest form of improvable reason manifested by Man." (Lyell, 1861: 505) By the fourth edition of 1873, the above paragraph had been cut, though there was still one reference to "conjecturing whether the successive steps in advance, by which a progressive scheme has been developed, may not admit of occasional strides constituting apparent breaks in an otherwise continuous series of psychical changes." (1873: 546) The terminology was more cautious and less overt, though in his *Journals*, Lyell struggled with the problems of continuity and saltationism even at later times.

[3]In his "Lecture on Evolution" (1876), an address to an American audience in New York, Huxley adduced three types of evidence for evolution: neutral, favourable, and demonstrative. The general fossil record provided no more than neutral evidence because of the gaps it contained, which if taken as absolute would constitute negative evidence against the hypothesis of the gradual evolution of species. The favourable evidence was provided by those cases where the fossil record contained evidence of intermediary forms between orders, such as *archoeopteryx* as the link between birds and reptiles. Demonstrative evidence was furnished by Marsh's work on the evolution of the horse, where a whole series of gradually evolving forms had been discovered. Evolution, as an inductive hypothesis, was thereby established by "the coincidence of the observed acts with theoretical requirements." (1876: 133)

[4]Huxley was not the only scientist in Victorian England to make this distinction between the adoption of a materialist terminology and the rejection of a materialist ontology. John Hughlings Jackson, the neurologist, made exactly the

same move. In neurology, he held that it was useless to use non-material terms to describe neurological disorders. But this acceptance of terminology did not license an extension of materialism to philosophy, where Hughlings Jackson adopted a form of monism related to that of Herbert Spencer. A more openly materialist stance was taken by John Tyndall, in his 1872 presidential address to the British Association for the Advancement of Science. It was he who coined the term "scientific materialism" to describe a special sort of materialism - one based on science and applicable within science. But on wider philosophical problems, such as the mind-body problem, he reverted to an agnosticism similar to that of Huxley. The term "scientific materialism," of which this appears to be the first use in the English-speaking world, was intended to distinguish the theory from "philosophical materialism," which in 19th century England had no open proponents among the prominent scientists and philosophers.

Chapter Four

[1]Wallace cited the following results of craniological studies for average brain sizes: "Teutonic family - 94 cubic inches, Esquimos: 91, Negroes: 85; Australians and Tasmanians: 82; Bushmen: 77" (1870: 188). He was astounded that Esquimos, whom he assumed to be less culturally and mentally developed than Europeans, had almost the same cranial capacity. He noted as well that the Neanderthal skull had a capacity of 75 cubic inches, and that the cave-man skull recently measured by Paul Broca was also quite large, though no exact figure was given. He rejected the view that brain size was irrelevant to mental ability, noting that skulls with circumferences of less than 19 inches and cranial capacities of 65 cubic inches were invariably associated with idiocy. He also claimed that great men - Napoleon and Cuvier were mentioned - "have always heads far above the average size." (1870: 190) Stephen Jay Gould in *Mismeasure of Man* (1981) criticized the 19th century practice of measuring cranial capacities and correlating them with intelligence quotients.

[2]Darwin was dismayed at his colleague's support for the supernatural, but despite this disagreement on a basic philosophical issue, maintained friendly personal and scientific relations with Wallace. An interesting critique from an unsympathetic source was contained in F. Engels' "Natural Science in the Spirit World," in his *Dialectics of Nature* (published 1933).

[3]In the same context, Wallace briefly considered an idealistic solution to the dilemma. Matter was known to humans only through its action as a force, and was understood on analogy to will power. Instead of having the three categories of

matter, force, and mind, Wallace wondered whether it might not be a "far simpler and more consistent belief that matter, as an entity distinct from force, does not exist; and that force is a product of mind." (1870: 369) However, he did not proceed any further with this speculation, suggested by the work of W. K. Clifford (1879).

[4]A graph is continuous at a point when the difference between the value of the graph at that point and its value close to that point tends to zero as the difference between the two points tends to zero. A graph is continuous when it is continuous at each point. The first derivative gives the rate of change of a graph. A continuous graph whose first derivative is continuous can be represented by a straight line, although a smooth curve also satisfies these two properties.

Chapter Five

[1]His writings included three volumes on comparative psychology: *Animal Intelligence* (1883), *Mental Evolutions in Animals* (1883), and *Mental Evolution in Man* (1888). His major work on Darwinism was a three volume work: *Darwin and After Darwin: An Exposition of the Darwinian Theory and Discussion of Post-Darwinian Questions* (1892-97), ed. by C. Lloyd Morgan: v.1. *The Darwinian Theory*, v.2. *Post-Darwinian Questions: Heredity and Utility*, and v.3. *Post-Darwinian Questions: Isolation and Physiological Selection.* Romanes' major philosophical work was *Mind and Motion and Monism*, (1895), ed. C. Lloyd Morgan. Romanes also produced a work on the theory of heredity, *An Examination of Weismannism* (1899).

[2]W. K. Clifford was a mathematician who died prematurely in 1879 at age 35. His theory of mind, including the concepts of "mind dust" and "ejective knowledge" is contained in "Mind and Body" (1874), in his *Lectures and Essays* (1879).

[3]The major critic of Romanes' evolutionary theory of mind was St. George Mivart, who in *Genesis of Species* (1871), had pointed out problems in Darwin's theory of evolution causing Darwin to devote a full chapter - the sixth - in the 5th edition of the *Origin of Species* to a reply. In *The Origin of Human Reason* (1889), Mivart criticized Romanes and proposed an alternate terminology and theory. What Mivart intended was to undo Romanes' intermediary step of the receptual linking the perceptual in animals and the conceptual in humans. Mivart claimed that lower recepts belonged to the sensitive stage of life, which he termed "consentional." In his view, Romanes' higher recepts and higher and lower concepts were purely conceptual. Mivart was a saltationist who argued for the discontinuity between brute and man, as well as between sensitive and non-sensitive life, and between living organisms and non-living matter.

Chapter Six

[1]Lloyd Morgan was born on Feb. 6, 1852 in London. He attended the Royal School of Mines, where he studied mining and metallurgy. Although his training would normally have led him to become a mining engineer or geologist, he never engaged in either profession. After graduation he did a post-graduate study year under T. H. Huxley. The year with Huxley had a profound influence on Lloyd Morgan, and led him to a career devoted to evolution both in its scientific and its philosophical aspects. In 1878 Lloyd Morgan was appointed lecturer at the Diocesan College at Rondebosch, South Africa. In 1883 he returned to England to teach at the University of Bristol, at first as lecturer, then as professor of geology and zoology, and finally as professor of psychology and ethics. From 1887-1909 he served as principal of the university, and upon his retirement in 1919 he was named emeritus professor. In 1899 Lloyd Morgan was elected to the Royal Society, the first fellow to be elected for work in psychology, and he was elected president of the psychological section of the British Association for the Advancement of Science at its annual meeting of 1921. Biographical material for Lloyd Morgan includes J. H. Parsons, "Conwy Lloyd Morgan: 1852-1936," in the *Obituary Notices of Fellows of the Royal Society*, Edwin Clarke "Morgan, Conwy Lloyd" in the *Dictionary of Scientific Biography*, T. A. Goudge, "Morgan, C. Lloyd," in the *Encyclopaedia of Philosophy*, as well as Lloyd Morgan's own "Autobiography" (1932), in *A History of Psychology in Autobiography*. Goudge's articles in the *Encyclopaedia of Philosophy* include "Emergentist Evolution," as well as articles on Darwin, Huxley, and Wallace.

[2]During this period Lloyd Morgan produced no less than 10 books: *Springs of Conduct: An Essay in Evolution* (1885), *Animal Biology: An Elementary Textbook* (1887), *Animal Life and Intelligence* (1890), *Animal Sketches* (1891), *An Introduction to Comparative Psychology* (1894), *Psychology for Teachers* (1894), *Habit and Instinct* (1896), *Animal Behavior* (1900), *Comparative Biology* (1905), and *The Interpretation of Nature* (1905). This prolific output appears to have been slowed by his assumption of administrative responsibilities at the university. His next book, *Instinct and Experience* (1912) was already part of his transitional period to emergent evolution.

[3]Richards discussed Lloyd Morgan in chapter 8, "Darwinism and the Demands of Metaphysics and Religion: Romanes, Mivart, and Morgan." He analyzed Lloyd Morgan's work in the 19th century context of the Romanes-Mivart debate on the scientific and philosophical status of mind, and centered on Lloyd Morgan's neutral monist philosophy of that time. There was only reference to

Lloyd Morgan's later work on emergentism, which was outside the time frame that Richards dealt with. (1987: 404)

[4]Lloyd Morgan's early writings on physical science were *Water and its Teachings in Chemistry, Physics and Physiography: A Suggestive Handbook* (1882), and *Facts Around Us: Simple Readings in Inorganic Science; with Experiments* (1884).

[5]The theory was developed at about the same time by Lloyd Morgan, James Baldwin, a philosopher, psychologist, and editor of the *Dictionary of Philosophy and Psychology* (1901), and H. F. Osborn, an American biologist and evolutionist. Lloyd Morgan discussed his theory, then called "indirect selection" based on "coincident variation" in *Habit and Instinct* (1896), and in an article in *Science*, Nov. 27, 1896; Osborn's article appeared in Science for April 3, 1896 and Nov. 27, 1896; Baldwin's in *Science*, March 20, 1896. Baldwin developed on this concept in *Mental Development in the Child and Race* (1896). Baldwin, Lloyd Morgan and Osborn discussed the concept in a joint article in *Science*, April 25, 1896. The concept was also discussed in "Organic Selection" in the *Dictionary of Philosophy and Psychology*, pp. 213-218, an article initialled by Baldwin, Lloyd Morgan, and several others. Though Lloyd Morgan did not use the concept in his later writings, Osborn continued to defend it in *Evolution and Religion in Education* (1926), where he considered it as a scientific concept linked to his own philosophy of creative evolution.

[6]Darwin used the example of the eye; this example is due to Stephen Jay Gould, "Not Necessarily a Wing" (1985), in *Natural History*, 94: 12-25, which discussed models of the intermediary stages of insect wings developed by Joel G. Kingsolver and M. A. R. Koehl. Their results were published in "Aerodynamics, Thermoregulation, and the Evolution of Insect Wings: Differential Scaling and Evolutionary Change" (1985), in *Evolution* 39: 488-504.

[7]An indication of the influence of Lewes on Lloyd Morgan can be seen by examining the title quotes Lloyd Morgan included for each part, chapter, and section in *Springs of Conduct*. Of the 17 persons quoted in this context, Lewes appeared most often, with 7 quotes out of a total of 34 quotes for all authors.

[8]Another instance of Lloyd Morgan's early use of the term "emerge" appeared in *Animal Life and Intelligence* (1890): "And in the brain, somehow associated with the explosion of its cells, consciousness and the mind-element emerges; of which we need only notice here that it belongs to a wholly different order of being from the physical activities and products with which we are at present concerned." (1890: 32)

[9]In turn Wallace reviewed Lloyd Morgan's *Animal Life and Intelligence* in *Nature*. The review was generally favorable, but not unexpectedly, Wallace objected

to Lloyd Morgan's monism, and the relation of identity Lloyd Morgan drew between psychosis and neurosis, which "implies that there can be no mind like that of man, or superior to it, without a brain formed of similar materials and similarly organized as the brain of man. This necessary connection, and even identity, of the two is, however, what is not proved, and not even, in my opinion, shown to be probable." (1891: 341)

[10]The *Monist* was a journal founded by Paul Carus in 1890, and from 1892 on subtitled "A Journal devoted to the Philosophy of Science." The complete series of articles by Lloyd Morgan comprised: "Mental Evolution, An Old Speculation in New Light" (1892, v.2), "The Doctrine of Auta" (1893, v.3), "Three Aspects of Monism" (1894, v.4), "Weismann on Heredity and Progress" (1894, v.4), "A Piece of Patchwork" (1895, v.5), "Naturalism" (1896, v.6), "Animal Automatism and Consciousness" (1896, v.7), "The Realities of Experience" (1897, v. 8), "Causation, Physical and Metaphysical" (1898, v.8), "The Philosophy of Evolution" (1898, v.8), "Vitalism" (1899, v.9), "Biology and Metaphysics" (1899, v.9), "Psychology and the Ego" (1900, v.10), and "The Conditions of Human Progress" (1900, v.10). Lloyd Morgan contributed the greatest number of articles after its editor to *Monist* during its first decade.

Chapter Seven

[1]The full title of Mill's work was *A System of Logic Ratiocinative and Inductive: Being a Connected View of the Principles of Evidence and the Methods of Scientific Investigation*. The full title more clearly indicated its goal of not merely serving as a formal logic, but also as a philosophy of science.

[2]The converse of this reasoning was applied in his ethical theory, where Mill formulated the "fallacy of composition": what was good for each individual did not necessarily add up to what was good for the society as a whole. This is discussed in *Logic* under Fallacies of Confusion (1872: 536-537)

[3]Lewes' work was published as five volumes in three series from 1874 to 1879 as follows: *Problems of Life and Mind*, 1st Series (1874, 1875), 2 volumes, *The Physical Basis of Mind*, 2nd series (1877), and *Problems of Life and Mind*, 3rd series (1879), 2 volumes. Like many Victorian works, those of Whewell and Mill in particular, Lewes dealt with the full breadth of knowledge and was encyclopaedic in scope. I will refer in what follows to each series by a Roman numeral, and volumes within a series by an Arabic numeral. The first series, subtitled *Foundations of a Creed* contained an introduction which discussed the method of science, its applications to metaphysics and the rules of philosophizing, and then dealt with a series of problems: the limitations of knowledge (I.1), the principles

of certitude, the relation of known to unknown, matter and force, force and cause, and the absolute in the correlation of feeling and motion (I.2). The second series, subtitled *The Physical Basis of Mind* dealt with the problems of the nature of life, the nervous mechanism, animal automatism, and the reflex theory (II). The third series, subtitled *The Study of Psychology: Its Object, Scope and Method*, dealt with the science of psychology and its relation to other sciences, notably biology and psychology (III.1), along with the further problems of mind as a function of the organism, the sphere of sense and logic of feelings, and the sphere of intellect and logic of signs (III.2). The third series was edited and published posthumously by George Elliot.

[4]Lewes had been a positivist and admirer of Comte. Lewes' *Biographical History of Philosophy* (1850) concluded with a chapter on Comte, and Lewes developed a distinct, though still positivist line of thought on life and mind in *Comte's Philosophy of the Sciences* (1853). In the chapter on the passage from the inorganic to the organic, he identified three laws governing the evolution of life. The first two laws were called "static laws," the third, a "dynamic law." Lewes referred to inorganic matter as "anorganic," complex but not yet vital compounds as "merorganic," and living matter as "teleorganic." The first law was: "The elements which compose Organic substances are the same as those which compose Inorganic substances; but in the Organic they occur as higher multiples." (1853: 145) The second was: "The presence of higher multiples is accompanied by an indefinite composition in lieu of a definite composition, and by a characteristic immediate synthesis of the elements." (1853: 152) The third was: "Merorganic substances become teleorganic by the assumption of a Spherical Form." (1853: 157) In this analysis, quantity played an essential role to distinguish the less complex inorganic from the more complex merorganic, and shape was the key to the distinction between the not-yet living merorganic and the vital teleorganic. The concept of "indefinite composition" mentioned in second law was not further explained, and on the face of it seemed to violate the principle of definite proportions of Dalton's atomic chemistry.

[5]Bergson developed this idea in "Le paralogisme psycho-physiologique" (1904). He returned to the problem of time and space in one of his last book length writings, *Duration and Simultaneity* (1921), where he attempted to show that his views of time and space were consonant with Einstein's theories of relativity.

[6]Though well received by some in England the volume met with opposition, especially from the traditional Darwinian and Spencerian evolutionists. A striking example was Hugh Elliot's *Modern Science and the Illusions of Professor Bergson* (1912), with a preface by Ray Lankester, a follower of Huxley and editor of his scientific works. This book was a protest against the introduction of what was

considered as illegitimate metaphysics into science, and rejected Bergson's philosophy as irrationalistic and anti-scientific.

[7]A Bergsonian influence was not unique to Lloyd Morgan among English writers on evolution. S. Herbert (a medical doctor) was the author of two popularizations of biological theory - *The First Principles of Heredity* (1908) and *The First Principles of Evolution* (1913). The use of the terminology "First Principles" suggested Spencer's volume introducing his *Synthetic Philosophy*, and this was no mere coincidence, as Herbert was a devoted Spencerian. His book on evolution, published in the same year as Lloyd Morgan's Spencer Lecture, was divided into three parts – inorganic, organic, and super-organic – again, based on Spencer's system. The first chapter set out Herbert's appreciation of Spencer, who was lauded for having "discovered a unifying principle of such far-reaching application, covering such various phenomena, to have given us the grand conception of a world embracing evolution." (1913: 3) In his penultimate chapter, Herbert recognized Spencer's law of the transformation of the homogeneous into the heterogeneous as the "formula of evolution." Yet the last chapter, on the philosophy of change, abruptly switched to a discussion of Bergson's creative evolution. Herbert admitted Bergson's critique of Spencer's philosophy of evolution as mechanistic, stating "If we have had before a science of evolution, Spencer's 'Synthetic Philosophy' not excluded, Bergson has given us for the first time a real philosophy of change." (1913: 318).

[8]The major writing of the "new realists" was *The New Realism: Cooperative Studies in Philosophy* (1912), co-authored by Edwin B. Holt, Walter T. Marvin, William P. Montague, Ralph Barton Perry, Walter B. Pitkin, and Edward Gleason Spaulding. This had been preceded by a programmatic statement "The Program and First Platform of Six Realists" (1910). Of these authors, both Marvin and Spaulding were explicitly referred to by Lloyd Morgan as influences, and Montague argued in the 1920s for a form of emergent materialism. Spaulding and Marvin were not alone among writers in the United States interested in the problem of novelty. William James in his posthumously published work, *Some Problems of Philosophy: A Beginning of an Introduction to Philosophy* (1911) devoted no less than half the book to this problem, with chapters such as IX: "The Problem of Novelty," X-XI: "Novelty and the Infinite," XII-XIII "Novelty and Causation." A second group of American philosophers issued *Essays in Critical Realism: A Co-operative Study of the Problem of Knowledge* (1920). This volume was co-authored by Durant Drake, Arthur O. Lovejoy, James B. Pratt, Arthur K. Rogers, George Santayana, Roy Wood Sellars, and C. A. Strong. Of these, Sellars developed his own monistic, naturalistic, and ultimately materialist version of emergent evolution under the term "evolutionary naturalism," and Lovejoy (1926)

wrote a commentary on emergent evolution.

9In making the distinction between additive and constitutive properties, Nernst followed the lead of the physical chemist Wilhelm Ostwald. This latter, in his *Outlines of General Chemistry* (1890), had in fact made a triple distinction, distinguishing between additive properties such as mass, where the simple addition rule applied; colligative properties, which did not change at all when components were combined to form compounds, such as the volume of gases; and constitutive properties, such as the boiling and melting points, which were not simple sums of the values of properties of the component parts.

10Spaulding also included a teleological notion in his theory and held that evolution was moving towards the realization of the extra-natural ideals of justice, truth, goodness, and beauty. The existence of these extra-natural ideals was linked to his notion of God as the sum-total of these ideals. God was both transcendent and immanent: transcendent insofar as the ideals were outside nature, immanent insofar as the ideals directed evolution towards their realization in the world.

11The term has been used previously by the 19th century psychologist and philosopher Wilhelm Wundt, and has been used by the Whiteheadian process philosopher Charles Hartshorne as well.

Chapter Eight

1*Instinct and Intelligence* followed up a discussion on that topic held at a joint meeting of the Aristotelian Society, British Psychological Society and the Mind Association in London in July, 1910. In his contribution to the symposium, Lloyd Morgan criticized the views of C. Wildon Carr, who defended Bergson's theory of intuition. Lloyd Morgan was in contact with Bergson in the autumn of 1912. He sent the French philosopher a copy of his *Instinct and Experience* just published that year, and Bergson replied to Lloyd Morgan in a letter of Nov. 21, 1912. (Archives 128/278) Bergson acknowledged receipt of the book, which he read "with interest" as a follow-up to Lloyd Morgan's *Habit and Intelligence.* Bergson noted that when he began to do philosophy, some 30 years earlier, he had been under the spell of the ideas of Herbert Spencer – "j'étais imbu des idées de Herbert Spencer." (1912: 2) He had then considered instinct as the outcome of a gradual process, though he soon changed his mind. Bergson continued: "It is possible that the method of Creative Evolution takes us, as you put it, 'outside or behind the biological field'" (1912: 3, my translation) But he considered the method important because it transcended gradualism, and suggested a non-mechanistic approach as an alternative.

[2]These notes are undated, and were catalogued under the identifying number 128/553 (box 128, document 553). They immediately follow another set of notes for a course of six lectures on "Psychology and Life," University of Bristol session 1912-1913, beginning Oct. 10, 1912. From this and other contextual evidence, I date the 219 pp. manuscript 128/553 as fall of 1912 as well. At the latest, it could not be later than 1913, when Lloyd Morgan's Spencer lecture "Spencer's Philosophy of Evolution" appeared, already featuring emergentist terminology.

Chapter Nine

[1]This was not the only time that Lloyd Morgan adopted multiple senses for a single term. In his Spencer lecture he distinguished three sorts of causation: causation as source, causation as ground, and causation as condition. Source was "a transcendent cause which produces the phenomena under consideration," ground was "the nature or constitution of that within which some process occurs," while condition referred to "some external influence" which set off or catalyzed an action or event. (1913: 24) For evolution, the cause as source was deity, the cause as ground was the collection of organisms that evolved, and the cause as condition was the changing milieu which formed the environment for the evolving organisms.

Chapter Ten

[1]In *System of Animate Nature* (1920), the biologist and popularizer of evolutionary theory, J. Arthur Thomson, distinguished three major stages of evolution, or becoming in nature, proposing distinct terms for each. The first was that of inorganic evolution, or "genesis," the second was phylogeny of species in the organic domain, or "evolution" proper, and the third, the evolution of humans through "history." Becoming was a continuous process marked by successive steps of "creative synthesis." The metaphor was that of a staircase: the evolutionary progress from inorganic to organic life and then to human life were "successive steps" on the staircase of evolution. Thomson's emphasis on the emergence of novelty in the context of continuous evolution was similar to Lloyd Morgan's point of departure in his 1912-15 period, and Thomson made numerous references to Lloyd Morgan's work. Thomson confirmed his adhesion to emergent evolution in the lectures entitled *Concerning Evolution* (1925) given at Yale University, the inaugural volume of the Terry Lecture series on natural theology: "There is something to be gained by considering what Professor Lloyd Morgan calls 'emergent evolution'. The whole ascent of life, not to speak of the genesis further

back still, is studded with puzzling 'emergences' - outcrops of genuine novelties." (1925: 205) Later Gifford lectures also featured discussions of emergence, including Michael Polanyi, *Personal Knowledge: Towards a Post-Critical Philosophy* (1958).

[2]Alexander capitalized each term referring to a level or category, and this practice will be adhered to when referring to terms of his system.

[3]The categories were listed by Alexander in "ranks" as follows: Identity, Diversity, and Existence; Universal, Particular, and Individual; Relation; Order; Substance, Causality and Reciprocity; Quantity and Intensity; Whole and Parts, Number; and Motion.

[4]Broad defended the paranormal in *Religion, Philosophy and Psychical Research* (1953), and *Lectures on Psychical Research* (1962).

[5]Wilson stated at the outset of *Sociobiology* (1975): "The higher properties of life are emergent. ...The recognition and study of emergent properties is holism, once a burning subject for philosophical discussion by such scientists as Lloyd Morgan (1922) and W. M. Wheeler (1927), but later, in the 1940s and 1950s, temporarily eclipsed by the triumphant reductionism of molecular biology. The new holism is much more quantitative in nature, supplanting the unaided intuition of the old theories with mathematical models. ... In the sections to follow we will examine several of the properties of societies that are emergent and hence deserving of a special language and treatment." (1975: 7) Yet Wilson has been accused of gross reductionism and biological determinism by critics such as R. C. Lewontin, Steven Rose, and Leon J. Kamin in *Not in our Genes: Biology, Ideology and Human Nature* (1984). The issue centers on whether or not Wilson in fact, not merely in statement, respected the autonomy of the social relative to the biological basis from which it emerged.

Chapter Eleven

[1]Other philosophers incorporated more restricted concepts of emergence in their metaphysical systems. C. E. M. Joad in his article "Emergence to Value" (1928) and his book *Matter, Life and Value* (1929) accepted the emergence of goodness, truth, and beauty at the mental level, though he rejected the emergence of mind from living matter, or life from inorganic matter. J. S. Haldane in *The Sciences and Philosophy* (1929), originally delivered as the Gifford Lectures of 1927-28, accepted a hierarchy of irreducible levels of interpretation in science, from the formal sciences of mathematics, to the factual ones of physics, chemistry, biology, and psychology, each interpretation adding something new to those that preceded it. Nonetheless, he rejected the view that what these sciences referred to –

matter, life, and mind – emerged from each other; rather they were there from the beginning of time. Alfred North Whitehead, Bertrand Russell's collaborator in the *Principia Mathematica*, was more sympathetic to emergence. In the preface to *Science and the Modern World* (1925: xi), he mentioned his debt to Lloyd Morgan and Alexander for their emphasis on creativity and novelty in the process of change. He reiterated his regard for Alexander's theory of space-time in *Religion in the Making* (1926: 111), and incorporated elements of creativity and novelty in the fundamental categories of his major work *Process and Reality* (1929), also delivered as Gifford Lectures. He termed creativity "the universal of universals," and the "ultimate principle" by which the many – the universe considered disjunctively – became, with attendant novelty, the one – the universe considered conjunctively. (1929: 31-32) Despite his accent on change, creativity, and novelty, however, Whitehead's system was not based on evolution, but the more abstract concept of "process," and his incorporation of ideas of emergence was secondary at best.

[2]Feibleman did not footnote *Ontology*, and so it is difficult to determine the influences on him of previous emergentists. But in "Theory of Integrative Levels," he referred to Alex Novikoff's 1945 article in *Science* and Ludwig von Bertalanffy's work in developmental biology and systems theory.

Chapter Twelve

[1]In 1926, the question of emergent evolution was discussed at a session of the VIth World Congress of Philosophy held at Harvard University, and at a meeting of the Aristotelian Society that same year. Interventions at the VIth World Congress of Philosophy on emergentism were those of Hans Driesch in "Emergent Evolution," H. Wildon Carr in "Life and Motion," A. O. Lovejoy in "The Meaning of 'Emergence' and its Modes," and W. M. Wheeler in "Emergent Evolution of the Social." Driesch attempted to combine his entelechy theory with emergentism in the concept of the superentelechy. The superentelechy's essence determined what was possible, while the actual was the result of its free will, this latter being an unpredictable emergent. As a result, evolution "would be half emergent and half non-emergent." (1926:8) Carr rejected emergence, while Lovejoy gave it qualified approval. Wheeler's views have been analyzed previously. The subject of emergent evolution was also debated at a meeting of the Aristotelian Society in 1926 by E. S. Russell, W. M. Morris and W. L. Mackenzie, all of whom, however, rejected it in favor of mechanistic or dualistic theories.

[2]In *Modern Materialism and Emergent Evolution*, McDougall examined Lloyd Morgan and Alexander's work, as well as other authors, some of whom cannot strictly speaking be termed emergentists, including L. T. Hobhouse in

Development and Purpose (1913) and C. A. Strong in *Origin of Consciousness* (1918). Hobhouse, though he defended the evolution of mind, as did Romanes and Darwin before him, did not develop any specifically emergentist theses related to qualitative novelty. Nor did Strong, whose theory was one of mind-body parallelism of a Leibnizian type. McDougall had cast his net too widely in his haste to condemn all forms of emergentism, including non-emergentist theories of mental evolution among his "catch" of emergentist ones. McDougall also examined Edmund Noble's *Purposive Evolution, the Link Between Science and Religion* (1926), which though it considered Lloyd Morgan's emergence, was principally concerned with a defence of divine purpose and teleology in evolution.

[3]Articles on emergence during this period included Reuben Ablowitz: "The Theory of Emergence" (1939), William Malisoff: "Emergence Without Mystery" (1939), W. T. Stace: "Novelty, Indeterminism, and Emergence" (1939), D. W. Gotshalk: "Causality and Emergence" (1942), Paul Henle: "The Status of Emergence" (1942), A. Campbell Garnett: "Scientific Method and the Concept of Emergence" (1942), Gustav Bergmann: "Holism, Historicism and Emergence" (1944), Orville T. Bailey: "Levels of Research in the Biological Sciences" (1945), Archie J. Bahm: "Emergence of Purpose" (1947), "Organic Unity and Emergence" (1947), and "Emergence of Values" (1948), Arthur Pap: "The Concept of Absolute Emergence" (1952), and C. W. Beruda: "On Emergence and Prediction" (1953).

Chapter Thirteen

[1]The full title of the work was *Foundations of the Unity of Science: Toward an International Encyclopaedia of Unified Science*, 2 vols. The editors were Otto Neurath, Rudolf Carnap, and Charles Morris.

[2]Needham also referred to The Brown-Conger debate as well as to works by the American biologist E. G. Conklin in *The Direction of Human Evolution* (1921) and the English anatomist, F. Wood-Jones in *Design and Purpose* (1942)

[3]The linking of emergentism and organicism was characteristic of many expositions of theories of integrative levels. Ritter and Bailey in "The Organismal Conception" (1928) were typical of this synthesis when they argued that emergent evolution and the organicist conception were complementary, the former in the analysis of the origin of organisms, the latter in analyzing their structure and function: "'Emergent evolution' is [the term used] when the origin and development of living things are of central interest, while the 'organismal conception' is [used] when their morphology and physiological functioning are considered." (1928: 334)

[4]Within the Marxist philosophical trend of thought, the emergentist problematic was dealt with, almost trivially, as the "law of the transformation of

quantity into quality." This was listed among of the axioms of the system, based on Hegel's concept, as set out in his writings on dialectical logic.

[5]Woodger translated Bertalanffy's *Modern Theories of Development* in 1933. Von Bertalanffy, in *Problems of Life: An Evaluation of Modern Biological and Scientific Thought* (1952), identified as distinct levels of organization those of atoms and molecules, cells and protoplasm, and individual organisms and supra-individual organizations, from small ecosystems to the whole of life on earth. In his subsequent work, however, von Bertalanffy dropped the concept of levels of organization in favour of that of open and closed systems, and argued that this latter was a scientific formulation preferable to the philosophical concept of emergence. For more on systems theory see von Bertalanffy's *General System Theory: Foundations, Development, Applications* (1968). Bertalanffy was the founder and editor of *General Systems: Yearbook of the Society for General Systems Research.*

[6]Huxley wrote the introduction to Teilhard's *The Phenomenon of Man* (1959). C. E. Raven in his very favourable intellectual biography of Teilhard, *Teilhard de Chardin: Scientist and Seer* (1962), devoted chapter seven to "Teilhard and emergent evolution," though the point seems to be that Raven was influenced by both Lloyd Morgan and Teilhard, rather than there having been a direct influence of Lloyd Morgan on Teilhard. Teilhard himself discussed the problems of complexity and novelty in evolution in "My Fundamental Vision," in *Toward the Future* (1975), ed. Rene Hague.

[7]In his preface to the second edition (1969) Kuhn stated: "Until a first version had been completed and largely revised, I anticipated that the manuscript would appear exclusively as a volume in the *Encyclopaedia of Unified Science.* The editors of that pioneering work had first solicited it, then held me firmly to my commitment, and finally waited with extraordinary tact and patience for a result. I am much indebted to them, particularly to Charles Morris, for wielding the essential goad and for advising me about the manuscript that resulted." (1969: 58) Kuhn's work was subsequently issued as a separate book and has had a tremendous influence on philosophy of science, though Kuhn himself has subsequently modified his views on "paradigm shifts," one of the key concepts in that work.

[8]Sperry's work was discussed at length at a Conference on Consciousness and Brain, held at Newport Beach, Calif. in 1973. The proceedings were published as *Consciousness and the Brain: A Scientific and Philosophical Inquiry*, ed. Gordon G. Globus, Grover Maxwell, and Irwin Savodnik. Sperry contributed an article entitled "Mental Phenomena as Causal Determinants in Brain Function," stressing his notion of "downward causation." Analyses and replies to Sperry included E. M. Dewan: "Consciousness as an Emergent Causal Agent in the Context of Control

System Theory," and William Wimsatt: "Reductionism, Levels of Organization and the Mind-Body Problem."

[9]Besides his important works on taxonomy in the context of the modern synthesis of evolution and genetics, Mayr has produced three major volumes on the history and philosophy of biology: *Evolution and the Diversity of Life: Selected Essays* (1976), *The Growth of Biological Thought: Diversity, Evolution and Inheritance* (1982), and *Towards a new Philosophy of Biology: Observations of an Evolutionist* (1988).

[10]Mayr noted in "From Molecules to Organic Diversity" (1964, in 1976) that he had made a similar point in a lecture in the "mid 1950s" about "the continuous emergence of new systems [that] often display characteristics that one could not have predicted on the basis of the properties of the unit elements." (1976: 68) According to Mayr, Neils Bohr, the quantum physicist, was in attendance and agreed that this occurred in the physical systems as well. "The properties of atoms, Bohr said, could not have been predicted in detail on the basis of a knowledge of isolated protons, neutrons and electrons," even though the atoms "owe their highly specific properties to the quantity and pattern of their simple unit components, the nuclei and electrons." (1977: 69)

[11]Salk's book appeared in the *Convergence Series* edited by Ruth Nanda Anshen, and as a popular exposition, did not contain footnotes or a bibliography indicating Salk's sources. Salk also contributed an earlier volume to the same series, *Man Unfolding* (1972).

[12]Salthe noted that the entities of the ecological and genealogical hierarchies were similar to David Hull's "interactors" and "replicators" respectively. The ecological or energy flow hierarchy extended from the molecule to the organism, population, ecosystem, biogeographic region, and the whole earth as a global ecosystem. The genealogical or reproductive hierarchy proceeded from genes through demes and species to monophyletic lineages, historical biota, and the total biosphere.

[13]Popper's views on evolution have themselves undergone an evolution. He was quite critical of Darwinism in *Poverty of Historicism* (1934, English translation 1957), arguing that the evolutionary hypothesis could not achieve the status of a scientific law because it was based on set of events possibly unique to the planet earth. The evolutionary hypothesis was similar to an historical statement, and no more. In "Natural Selection and the Emergence of Mind" (1978), Popper belatedly realized that Darwin never intended natural selection to be the sole factor of evolution, and as a result did not consider it identical with evolution. As one among several factors of evolution, natural selection was both testable and refutable, and so had the status of a scientific theory, according to

Popper's refutation criterion. In fact, at least one part of Darwin's work, pangenesis as a theory of heredity, had already been refuted.

[14]Bunge cited as early influences Needham, Novikoff, and Schneirla. He also mentioned Hartmann's *Neue Wege der Ontologie*, as well as Alexander's *Space, Time and Deity*, Lloyd Morgan's *Emergence of Novelty*, and Roy Wood Sellars' *Evolutionary Naturalism.*

[15]Bunge's views on systems were developed in articles which initially appeared in *International Journal of General Systems*, including "Things" (1974), "The GST Challenge to the Classical Philosophies of Science" (1977), "A Theory of Properties and Kinds" (1977, with A. Sangalli), and "Analogy Between Systems" (1981). The systems point of view was discussed at length in vol. 4 of the *Treatise on Basic Philosophy*, which was subtitled "A World of Systems." Archie J. Bahm has written an article, "Five Systems Concepts of Society" (1983) which commented on Bunge's distinction between atomism, systemism, and holism, including the further concepts of structuralism and organicism. A similar point was also made by Bahm in "Five Types of System Philosophy" (1981).

[16]Other of Bunge's writings on levels include "Do the Levels of Science Reflect the Levels of Being?" in *Metascientific Queries* (1959), "Levels: A Semantic Inquiry" (1960), reprinted with modifications as "Levels" in *The Myth of Simplicity: Problems of Scientific Philosophy* (1963), "On The Connections Among Levels" (1960), and "The Metaphysics, Epistemology and Methodology of Levels" (1969), reprinted with modifications in *Method, Models and Matter* (1973).

[17]Bunge's writings on the mind-body problem include "Emergence and the Mind" (1977), "The Mind-Body Problem in an Evolutionary Perspective" (1978), "From Neuron to Behavior and Mentation: An Exercise in Levelmanship" (1980), "The Psychoneural Identity Theory" (1980), as well as his books *The Mind-Body Problem: A Psychobiological Approach* (1980), and *Philosophy of Psychology* (1987, with R. Ardilla).

[18]The debate over emergentism was a central feature of the series of articles which appeared in the review *Neuroscience*. Initiated by Bunge's "Emergence and the Mind" (1977), the debate eventually included Roger Sperry: "Mind-Brain Interaction: Mentalism, Yes; Dualism, No" (1980), Donald Mackay: "Selves and Brains" (1978), J. J. C. Smart: "Physicalism and Emergence" (1981), D. O. Hebb: "Consider Mind as a Biological Phenomenon" (1981), and Patricia Smith-Churchland: "Mind-Brain Reduction: New Light from the Philosophy of Science" (1982).

BIBLIOGRAPHY

Ablowitz, Reuben (1939). "The Theory of Emergence", *Phil. Sc.* 6: 1-16

Alexander, Samuel (1920). *Space, Time and Deity*. Gifford Lectures for 1916-18, 2 vols. (London: Macmillan and Co.)

Aristotle. *The Complete Works of Aristotle*. Revised Oxford translation, ed. Jonathan Barnes. (Princeton, N.J. : Princeton University Press)

Ayala, Francisco and Th. Dobzhansky (1974). *Studies in the Philosophy of Biology: Reductionism and Related Problems* (Berkeley: Univ. of California Press)

Bagehot, Walter (1873). *Physics and Politics, or Thoughts on the Application of the Principles of "Natural Selection" and "Inheritance" to Political Society* (London: Henry S. King and Co.)

Bahm, Archie J. (1947a). "Emergence of Purpose", *J. Phil.* 44: 633-636

Bahm, Archie J. (1947b). "Organic Unity and Emergence", *J. Phil.* 44: 241-244

Bahm, Archie J. (1948). "Emergence of Values", *J. Phil.* 45: 411-414

Bahm, Archie J. (1954). "Evolutionary Naturalism", *Phil. and Phenom. Res.* 15: 1- 12 [A Symposium in Honor of Roy Wood Sellars]

Bahm, Archie J. (1981). "Five Types of System Philosophy", *Int. J. Syst. Th.* 6: 233-237

Bahm, Archie J. (1983). "Five Systems Concepts of Society", *Gen. Syst.* 44-57

Bailey, Orville T. (1945). "Levels of Research in the Biological Sciences", *Phil. Sc.* 12: 1- 7

Baldwin, James Mark, ed. (1901) *Dictionary of Philosophy and Psychology*, 4 vols. (NY The Macmillan Company)

Baldwin, James, Conwy Lloyd Morgan, and H. F. Osborn (1896) "Organic Selection", *Science*, April 25, 1896.

Barker, Eileen (1976). "Apes and Angels: Reductionism, Selection and Emergence in the Study of Man", *Inquiry* 19: 367 - 399

Barrett, Paul H. (ed. 1974). *Metaphysics, Materialism and the Evolution of Mind: Early Writings of Charles Darwin* (Chicago: Univ. of Chicago Press)

Baylis, Charles A. (1929). "The Philosophic Functions of Emergence", *Phil. Rev.* 28: 372- 384

Beckner, Morton (1974). "Reduction, Hierarchies and Organicism", in Ayala, Francisco and Th. Dobzhansky (1974). *Studies in the Philosophy of Biology: Reductionism and Related Problems* (Berkeley: Univ. of California Press)

Bergmann, Gustav (1944). "Holism, Historicism and Emergence", *Phil. Sc.* 11: 209 - 221

Bergson, Henri (1889). *Time and Free Will*, tr. F. L. Pogson 1910 (London: George Allen and Unwin, 1950).

Bergson, Henri (1896). *Matter and Memory*, tr. N. M. Paul and W. Scott Palmer, 1911 (London: George Allen and Unwin, 1978)

Bergson, Henri (1907). *Creative Evolution*, tr. Arthur Miller, 1911 (NY: Modern Library, 1940)

Bergson, Henri (1922) *Duration and simultaneity, with Reference to Einstein's Theory*, tr. Leon Jacobson, with an intro. by Herbert Dingle (Indianapolis: Bobbs-Merrill, 1965)

Bertalanffy, Ludwig von (1933) *Modern Theories of Development; an Introduction to Theoretical Biology*, trans. and adapted by J.H. Woodger (London: Oxford University Press)

Bertalanffy, Ludwig von (1952). *Problems of Life: an Evaluation of Modern Biological Thought* (London, Watts)

Bertalanffy, Ludwig von (1968). *General System Theory: Foundations, Development, Applications* (NY: G. Braziller)

Beruda, C. W. (1953). "On Emergence and Prediction", *J. Phil.* 50: 269-273

Boodin, John Elof (1925). *Cosmic Evolution: Outlines of Cosmic Idealism* (New York: The Macmillan Co.)

Bowler, Peter J. (1967). "The Changing Meaning of Evolution", *J. Hist Ideas* 37: 95-114

Bowler, Peter J. (1983). *The Eclipse of Darwinism: Anti-Darwinian Evolution Theories in the Decades Around 1900* (Baltimore: Johns Hopkins Press)

Bowler, Peter J. (1984). *Evolution: The History of an Idea.* (Berkeley: U. of California Press)

Bowler, Peter J. (1986). *Theories of Human Evolution: A Century of Debate, 1844-1944.* (Baltimore: The Johns Hopkins Press)

Brackman, Arnold C. (1980). *A Delicate Arrangement: The Strange Case of Charles Darwin and Alfred Russel Wallace* (NY: Times Books)

Broad, C. D. (1926). *Mind and Its Place in Nature* (London: Routledge and Kegan Paul)

Broad, C.D. (1953). *Religion, Philosophy, and Psychical Research: Selected Essays* (London: Routledge & Kegan Paul)

Broad, C.D. (1962). *Lectures on Psychical Research* (London, Routledge & Kegan Paul)

Brown, Harald Chapman (1925). "The Material World - Snark of Boojum", *J. Phil.* 22: 197-214

Brown, Harald Chapman (1926). "A Materialist's View of the Concept of Levels", *J. Phil.* 23: 113-120

Brown, Harald Chapman (1933). "Mind - An Event in Physical Nature", *Phil. Rev.* 42: 130-155

Bunge, Mario (1959a). *Causality, the Place of the Causal Principle in Modern Science* (Cleveland: World Publishing Co.)

Bunge, Mario (1959b). *Metascientific Queries* (Springfield, Ill: Charles C. Thomas)

Bunge, Mario (1960a). "Levels: A Semantic Preliminary", *Rev. Meta.* 13: 396-406

Bunge, Mario (1960b). "On the Connections Among Levels", in *Proc. XIIth Int. Cong. Phil.* 6: 63-70

Bunge, Mario (1963). *The Myth of Simplicity: Problems of Scientific Philosophy* (Englewood Cliffs, NJ: Prentice Hall)

Bunge, Mario (1967). *Scientific Research I: The Search for System* (Heidelberg: Springer-Verlag)

Bunge, Mario (1969). "The Metaphysics, Epistemology and Methodology of Levels", in *Hierarchical Structures*, ed. L. L. Whyte, A. G. Wilson and D. Wilson (NY: American Excelsior)

Bunge, Mario (1973). *Method, Model and Matter* (Dordrecht: D. Reidel Pub. Co.)

Bunge, Mario (1973 - 89). *Treatise on Basic Philosophy* (Dordrecht: D. Reidel Pub. Co.):

v. 1. Semantics I: Sense and Reference (1973)
v. 2. Semantics II: Interpretation and Truth (1974)
v. 3. Ontology I: The Furniture of the World (1977)
v. 4. Ontology II: A World of Systems (1979)
v. 5. Epistemology and Methodology I: Exploring the World (1983)
v. 6. Epistemology and Methodology II: Understanding the World (1983)
v. 7. Epistemology and Methodology III: Philosophy of Science and Technology, Parts I and II (1985)
v. 8. Ethics and the Good Life (1989)

Bunge, Mario (1974a). "Metaphysics and Science", *Gen. Syst.* 19: 15-18

Bunge, Mario (1974b). "Things", *Int. J. of Gen.Syst.* 1: 229-236

Bunge, Mario (1977a). "Emergence and the Mind", *Neurosci.* 2: 501-509.

Bunge, Mario (1977b). "General Systems and Holism", *Gen. Syst.* 22: 87-90

Bunge, Mario (1977c). "Levels and Reduction", *Am. J. Physio.* 233 (R3): 75-82

Bunge, Mario (1977d). "The GST challenge to the classical philosophies of science". *Int. J. of Gen.Syst.* 4: 29-376

Bunge, Mario (1978). "The Mind-Body Problem in an Evolutionary Perspective", in *Mind and Brain*, CIBA Foundation Symposium 69 (N. S.), (NY: Excerpta Medica)

Bunge, Mario (1980a). "From Neuron to Behavior and Mentation: An Exercise in Levelmanship", in *Information Processing in the Nervous System*, ed. H. M. Pinker and W. D. Willis (NY: Raven Press)

Bunge, Mario (1980b) "The Psychoneural Identity Theory", in Dalbir Bindra, ed., *The Brain's Mind*, (New York: Gardner Press), pp. 89-108

Bunge, Mario (1980c). *The Mind-Body Problem, a Psychobiological Approach* (Oxford: Pergamon Press)

Bunge, Mario (1981). "Analogy between systems", *Int. J. of Gen. Syst.* 7: 221-223

Bunge, Mario (1982). *Scientific Materialism* (Oxford: Pergamon Press)

Bunge, Mario and Ruben Ardila (1987). *Philosophy of Psychology* (Berlin: Springer Verlag)

Campbell, Donald C. (1974). " 'Downward Causation' in Hierarchically Organized Biological Systems, in Ayala, Francisco and Th. Dobzhansky (1974). *Studies in the Philosophy of Biology: Reductionism and Related Problems* (Berkeley: Univ. of California Press)

Carnap, Rudolf (1928) *The Logical Structure of the World*, tr. Rolf A. George (Berkeley: University of California Press, 1967)

Carnap, Rudolf (1963). "Reply", in Schilpp, Paul Arthur (1963) *The Philosophy of Rudolf Carnap*, The Library of Living Philosophers, v.11 (La Salle, Ill., Open Court)

Carr, H. Wildon. (1926). "Life and Matter", in Brightman, Edgar S. (ed.) *Proceedings of the Sixth International Congress of Philosophy* (NY: Longmans, Green and Co.), pp. 9-19

Chambers, Robert (1844). *Vestiges of the Natural History of Creation* (NY: Wiley and Putnam, 1846)

Chambers, Robert (1845). *Explanations: A Sequel to Vestiges of the Natural History of Creation* (NY: Wiley and Putnam, 1846)

Churchland, Patricia. (1981). "Mind-Brain Reduction: New Light from the Philosophy of Science", *Neurosci.* 7: 1041-1047

Clarke, Edwin (1970). "Morgan, Conwy Lloyd", *The Dictionary of Scientific Biography*, ed. C. C. Gillespie, vol. for M: 512-513

Clifford, William Kingdon (1879) *Lectures and Essays*, 2 vols., edited by Leslie Stephen and Frederick Pollock (London, Macmillan)

Cohen, I. Bernard, ed. (1958). *Isaac Newton's Papers and Letters on Natural Philosophy, and Related Documents* (Cambridge, Mass: Harvard Univ. Press)

Collins, F. Howard (1889). *An Epitome of The Synthetic Philosophy*, with a preface by Herbert Spencer. (London, Williams and Norgate)

Comte, Auguste, (1830-35). *The Positive Philosophy*, translated and condensed by Harriet Martineau (New York: D. Appleton and Co, 1853)

Conger, George P. (1921). "Evolution and Epitomization", *Monist* 21: 536-584

Conger, George P. (1925). "The Doctrine of Levels", *J. Phil.* 22: 309- 321

Conger, George P. (1926). "What are the Criteria of Levels", *J. Phil.* 23: 589 - 598

Conger, George P. (1927). "Review of Smuts' Holism and Evolution", *J. Phil.* 24: 136-137

Conger, George P. (1928). "A Hypothesis of Realms", *J. Phil.* 25: 205 - 217

Conger, George P. (1929). *New Views of Evolution* (New York: The Macmillan Co.)

Conger, George P. (1931). *A World of Epitomizations: A Study in the Philosophy of the Sciences* (Princeton: Princeton University Press)

Conger, George P. (1933). "Epitomization and Epistemology", *Monist* 33: 73-87

Conklin, Edwin Grant, (1921) *The Direction of Human Evolution* (NY: Scribner)

Darwin, Charles (1859). *On the Origin of Species by Means of Natural Selection, or, the Preservation of Favoured Races in the Struggle for Life*. A facsimile of the First Edition (Cambridge, Mass: Harvard Univ. Press, 1964)

Darwin, Charles (1862). *On the Various Contrivances by which British and Foreign Orchids are Fertilised by Insects, and on the Good Effects of Intercrossing* (London: John Murray)

Darwin, Charles (1868). *On the Variation of Animals and Plants under Domestication*, 2 vols, 2nd edition 1875 (NY: D. Appleton and Co.)

Darwin, Charles (1870-71). *The Descent of Man, and Selection in Relation to Sex*, 2 vols. 2nd edition 1874 (NY: Macmillan and Co. 1902)

Darwin, Charles (1872), *The Origin of Species by Means of Natural Selection, or, the Preservation of Favoured Races in the Struggle for Life*. 6th edition (NY: P. F. Collier and Son, 1902)

Darwin, Charles (1890). *The Expression of Emotions in Man and Animals*, 2 vols. 2nd edition, ed. Francis Darwin (NY: D. Appleton and Co., 1897)

Darwin, Charles (1894-98). Charles Darwin's Works (NY: D. Appleton and Co.), 18 vols:

v. 1. *Journal of Researches during the Voyage of H. M. S. Beagle*

v. 2. *The Structure and Distribution of Coral Reefs*

v. 3. *Geological Observations of Volcanic Islands and Part of South America*

v. 4-5. *Origin of Species by Means of Natural Selection*

v. 6-7. *The Variation of Animals and Plans under Domestication*

v. 8-9 *The Descent of Man and Selection in Relation to Sex*

v. 10. *The Expression of Emotions in Man and Animals*
v. 11. *The Movements and Habits of Climbing Plants*
v. 12. *Insectivorous Plants*
v. 13. *The Effects of Cross and Self-Fertilization in the Vegetable Kingdom*
v. 14. *The Different Forms of Flowers on Plants of the Same Species*
v. 15 *The Power of Movement in Plants*
v. 16. *The Formation of Vegetable Mold Through the Action of Worms*
v. 17-18 *The Life and Letters of Charles Darwin (ed. Francis Darwin)*

Darwin, Charles (1958). *Charles Darwin's Essay of 1844*, in *Evolution by Natural Selection*, with a foreword by Gavin de Beer (Cambridge, University Press, 1958)

Darwin, Charles (1960). *Darwin's Notebooks on Transmutation of Species*, edited and with an introduction and notes by Gavin de Beer (1960). *Bull. Br. Mus. (Nat. Hist.)*, Part I, 2: 27-73, Part II, 2: 75-118, Part III, 2: 119-150, Part IV: 2: 151-183, Part V (Addenda and Corrigenda) 2: 185-200, Part VI (Pages excised by Darwin), 3: 129-176

Darwin, Charles (1974), and Thomas Henry Huxley. *Autobiographies*, edited and with an introduction by Gavin de Beer (London: Oxford University Press)

Darwin, Charles (1975). *Charles Darwin's Natural Selection, Being the Second Part of his Big Species Book written from 1856 to 1858.* edited from manuscript by R. C. Stauffer (Cambridge: University Press, 1975)

Darwin, Charles (1977) *"M' and "N" Notebooks*, in *Metaphysics, Materialism and the Evolution of Mind*, transcribed and annotated by Paul H. Barrett (Chicago: U. of Chicago Press).

Drake, Durant et. al (1920). *Essays in Critical Realism: a Co-operative Study of the Problem of Knowledge* (London, Macmillan and Co.)

Driesch, Hans (1908). *The Science and Philosophy of the Organism* (London: A. & C. Black)

Driesch, Hans. (1926). "Emergent Evolution", in Brightman, Edgar S. (ed.) *Proceedings of the Sixth International Congress of Philosophy* (NY: Longmans, Green and Co.), pp. 1-9

Drummond, Henry (1887). *The Ascent of Man* (NY: James Pott and Co., 1911)

Dudycha, George J. (1929). "What is Evolution", *Sc. Monthly* 29: 317 - 332

Eiseley, Loren (1979). *Darwin and the Mysterious Mr. X: New Light on the Evolutionists* (NY: E. P. Dutton)

Eldredge, Niles (1979). "Alternative Approaches to Evolutionary Theory", in *Bull. Carnegie Mus. Nat. Hist.* 13: 7-19

Eldredge, Niles and Stephen Jay Gould. (1972). "Punctuated Equilibria: An Alternative to Phyletic Gradualism", in *Models in Paleobiology, ed. Thomas J. M. Schopf* (San Francisco: Freeman, Cooper and Co.)

Eldredge, Niles and Stanley N. Salthe (1984). "Hierarchy and Evolution", in *Oxford Surveys in Evolutionary Biology*, ed. R. Dawkins and M. Ridley, 1: 184-208

Eliott, Hugh (1912). *Modern Science and the Illusions of Professor Bergson* (1912), with a preface by Ray Lankester (London: Macmillan and Company)

Emerson, A. E. (1942). "Basic Comparisons of Human and Insect Societies", in R. Redfield, ed. *Levels of Integration in Biological and Social Systems* (Lancaster: J. Cattell Press)

Evans, L. T. (1984). "Darwin's Use of the Analogy between Artificial and Natural Selection", *J. Hist. Bio.* 17: 113-140

Feibleman, James K. (1951). *Ontology* (Baltimore: Johns Hopkins Press)

Feibleman, James K. (1954). "Theory of Integrative Levels", *Br. J. Phil. Sc.* 5: 59-66

Feigl, Herbert (1958). "The Mental and the Physical", in *Minnesota Studies in the Philosophy of Science: Concepts, Theories, and the Mind-Body Problem*, ed. by H. Feigl, M. Scriven, and G. Maxwell. (Minneapolis, Minn: University of Minnesota Press)

Feigl, Herbert (1963). "Physicalism, Unity of Science and the Foundations of Psychology", in Schilpp, Paul Arthur (1963) *The Philosophy of Rudolf Carnap*, The Library of Living Philosophers, v.11 (La Salle, Ill., Open Court)

Garnett, A. Campbell. (1942). "Scientific Method and the Concept of Emergence", *J. Phil.* 39: 477- 486

Gerard, R. W. (1942). "Higher Levels of Integration", in R. Redfield, ed. *Levels of Integration in Biological and Social Systems* (Lancaster: J. Cattell Press)

Gerard, R. W. and A. E. Emerson (1945). "Extrapolation from the Biological to the Social", *Science* 67: 583-584

Ghiselin, Michael T. (1969). *The Triumph of the Darwinian Method* (Chicago: Univ. of Chicago Press)

Ghiselin, Michael T. (1987). "Species Concepts, Individuality and Objectivity", *Bio. and Phil.* 2: 127-144

Gilson, E. (1926). "Rapport sur le VIe Congrès International de Philosophie", Division A: "L'hypothèse de l'évolution émergente, sa signification et l'état présente de la controverse", *Revue de Meta. et de Morale*, 33: 516-518

Globus, Gordon G. and Grover Maxwell, and Irwin Savodnik (eds. 1976). *Consciousness and the Brain: A Scientific and Philosophical Inquiry* (NY: Plenum Press)

Goldschmidt, Richard Benedict (1940). *The Material Basis of Evolution* (New Haven, Yale University Press)

Gotshalk, D. W. (1935). "Some Problems of Evolution" *Monist* 45: 80-95

Gotshalk, D. W. (1942). "Causality and Emergence", *Phil. Rev.* 51: 397-405

Goudge, T. A. (1965). "Another Look at Emergent Evolutionism", *Dialogue* 4: 273-285

Goudge, T. A. (1967a). "Emergentist Evolution", in *Encyclopaedia of Philosophy*, ed. Paul Edwards (NY: Macmillan and Co.)

Goudge, T. A. (1967b). "Darwin, Charles", "Huxley, Thomas Henry" and "Wallace, Alfred Russel" in *Encyclopaedia of Philosophy*, ed. Paul Edwards (NY: Macmillan and Co.)

Goudge, T. A. (1973). "Evolutionism", in *Dictionary of the History of Ideas: Studies of Selected Pivotal Ideas*, ed. Philip P. Wiener (NY. Scribner's and Sons), 2: 174-189

Gould, Stephen J. (1977a). *Every Since Darwin: Reflections in Natural History* (NY: W. W. Norton)

Gould, Stephen J. (1977b). *Ontogeny and Phylogeny* (Cambridge, Mass: Harvard Univ. Press)

Gould, Stephen J. (1980). "Is a New and General Theory of Evolution Emerging?", *Paleobio.* 6: 119-130

Gould, Stephen Jay. (1981) *The Mismeasure of Man* (NY: W. W. Norton)

Gould, Stephen Jay (1985) "Not Necessarily a Wing", *Natural History*, 94: 12-25

Gould, Stephen J. and Niles Eldredge (1977). "Punctuated Equilibria: the Tempo and Mode of Evolution Reconsidered", *Paleobio.* 3: 115-151

Gray, Asa (1876). *Darwiniana: Essays and Reviews Pertaining to Darwinism*, ed. A. Hunter Dupress (Cambridge, Mass: The Belknap Press of the Harvard University Press)

Greene, John C. (1959). *The Death of Adam: Evolution and Its Impact on Western Thought* (Ames, Iowa: Iowa State Univ. Press)

Greene, John C. (1961). *Darwin and the Modern World View* (Toronto: Mentor Books)

Greene, John C. (1975). "Reflections on the Progress of Darwin Studies", *J. Hist. Bio.* 10: 1-27

Greene, John C. (1981). *Science, Ideology and World View: Essays in the History of Evolutionary Ideas* (Berkeley: Univ. of California Press)

Grinnell, George James (1985). "The Rise and Fall of Darwin's Second Theory", *J. Hist. Bio.* 18: 51-70

Gruber, Howard E. (1981). *Darwin on Man: A Psychological Study of Scientific Creativity* (Chicago: U. of Chicago Press)

Gruenberg, Benjamin C. (1930). *The Story of Evolution: Facts and Theories on the Development of Life* (London: Chapman and Hall)

Gutting, Gary, ed (1980). *Paradigms and Revolutions: Applications and Appraisals of Thomas Kuhn's Philosophy of Science* (Notre Dame and London: University of Notre Dame Press)

Habermeal, Lawrence L. (1967). *Value in the Evolutionary World Views of Samuel Alexander, C. Lloyd Morgan, and Pierre Teilhard de Chardin* (Unpublished doctoral dissertation, Boston University)

Haeckel, Ernst (1892). "Our Monism: The Principles of a Consistent, Unitary World-View", *Monist*, 4: 4811-486

Haeckel, Ernst (1899). *The Riddle of the Universe at the Close of the Nineteenth Century* (NY: Harper and Brothers Publishers)

Haldane, J. S. (1929). *The Sciences and Philosophy*. Gifford Lectures for 1927-28 (NY: Doubleday)

Hartmann, Nicolai (1951). *New Views of Ontology* (NY: The Macmillan Co.)

Hebb, D. O. (1981). "Consider Mind as a Biological Problem" *Neurosci.* 6: 2419-2422

Henle, Paul. (1942). "The Status of Emergence", *J. Phil.* 39: 486-492

Herbert, S. (1908). *The First Principles of Heredity* (London: Adam and Charles Black)

Herbert, S. (1913). *The First Principles of Evolution* (London: Adam and Charles Black)

Herrick, C. Judson (1949). "A Biological Survey of Integrative Levels", in *Philosophy of the Future: The Quest of Modern Materialism*, ed. R. W. Sellars, V. J. McGill, Marvin Farber (NY: The Macmillan Co.)

Hodge, M. J. S. (1972). "The Universal Gestation of Nature: Chambers' Vestiges and Explanations", *J. Hist. Bio.* 5: 127-151

Holt, Edwin B. et al (1912). *The New Realism: Co-operative Studies in Philosophy* (New York, Macmillan).

Hume, David (1779). *Dialogues Concerning Natural Religion*, edited with an introd. by Henry D. Aiken. (NY: Hafner Pub. Co, 1951)

Huxley, Julian and Huxley, T. H. (1947). *Evolution and Ethics: 1893-1943* (London: The Pilot Press)

Huxley, Thomas Henry (1854). "Review of Vestiges of the Natural History of Creation", in *Br. and Foreign Medico. Chir. Rev.* 13: 332-343

Huxley, Thomas Henry (1860). "The Origin of Species", in *Coll. Essays*, 2: 22-79

Huxley, Thomas Henry (1862). *On the Origin of Species, or, The Causes of the Phenomena of Organic Nature* (New York: D. Appleton and Company, 1869)

Huxley, Thomas Henry (1863). Man's Place in Nature, in *Coll. Essays*, 1: 1-208

Huxley, Thomas Henry (1868). "On the Physical Basis of Life", in *Coll. Essays*, 1: 130-165

Huxley, Thomas Henry (1870). "On Descartes' 'Discourse'", in *Coll. Essays*, 1: 166-198

Huxley, Thomas Henry (1874). "On the Hypothesis that Animals are Automata, and its History", in *Coll. Essays*, 1: 199-250

Huxley, Thomas Henry (1876). "Lectures on Evolution", in *Coll. Essays*, 4: 46-138

Huxley, Thomas Henry (1893). "Evolution and Ethics", in *Coll. Essays*, 9: 46-116

Huxley, Thomas Henry (1893-94). *Collected Essays* (NY: D.Appleton and Co.), 9 vols:

v.1. *Methods and Results*

v.2. *Darwiniana*

v.3. *Science and Education*

v.4. *Science and the Hebrew Tradition*

v.5. *Science and the Christian Tradition*

v.6. *Hume and Helps to the Study of Berkeley*

v.7. *Man's Place in Nature and Other Anthropological Essays*

v.8. *Discourses, Biological and Geological*

v.9. *Evolution and Ethics, and Other Essays*

Huxley, Thomas Henry (1903). *Life and Letters of Thomas Henry Huxley*, ed. Leonard Huxley, 3 vols. (London: Macmillan and Co.)

James, William (1911). *Some Problems of Philosophy; a Beginning of an Introduction to Philosophy* (NY: Longmans, Green, and Co.)

Jennings, H. S.(1927). "Diverse Doctrines of Evolution, Their Relation to the Practice of Science and of Life", *Science* 65: 19-25

Jerison, Harry J. (1973). E*volution of the Brain and Intelligence* (NY: Academic Press)

Joad, C. E. M. (1928). "Emergence to Value", *Proc. Aristot. Soc.* 28: 71-96

Joad, C. E. M. (1929). *Matter, Life and Mind* (London: Macmillan and Co.)

Keekes, John (1966). "Physicalism, The Identity Theory, and the Doctrine of Emergence", *Phil. Sc.* 33: 360-375

Kim, Jaegwon (1978). "Supervenience and Nomological Incommensurables", *Am. Phil. Quart.* 15: 149-156

Kim, Jaegwon (1982). "Psychophysical Supervenience", *Phil. Stud.* 41: 51-70

Kingsolver, Joel G. and M. Koehl (1985). "Aerodynamics, Thermoregulation, and the Evolution of Insect Wings: Differential Scaling and Evolutionary Change", *Evolution* 39: 488-504

Kohn, David, ed. (1985). *The Darwinian Heritage* (Princeton: Princeton Univ. Press in association with Nova Pacifica)

Koestler, Arthur and J. R. Smythies (1969). *Beyond Reductionism: New Perspectives in the Life Sciences* (NY: Macmillan)

Kropotkin, Petr Alekseevich (1902). *Mutual Aid: a Factor of Evolution* (NY: McClure, Philips & Co.)

Kuhn, Thomas (1969). *The Structure of Scientific Revolutions* (Chicago: Univ. of Chicago Press)

Kuhn, Thomas (1977). *The Essential Tension: Selected Studies in Scientific Tradition and Change* (Chicago: Univ. of Chicago Press)

Lamarck, Jean Baptiste (1809). *Zoological Philosophy: An Exposition with Regard to the Natural History of Animals*, trans. with an introduction by Hugh Elliot (NY: Hafner Publishing Co, 1963)

La Mettrie, Julien (1747). *Man a Machine*, tr. with philosophical and historical notes by Gertrude Carman Bussey (LaSalle, Illinois: Open Court, 1961)

Le Conte, Joseph (1874). "Correlation of the Vital with Chemical and Physical Forces", in Balfour Stewart (1874). *The Conservation of Energy* (London: Kegan Paul)

Le Conte, Joseph (1891). *Evolution: Its Nature, Its Evidences and Its Relation to Religious Thought* (NY: D. Appleton and Co., 1899)

Leibniz, G. W. (1981). *New essays on Human Understanding*. tr. and ed. by Peter Remnant and Jonathan Bennett. (NY: Cambridge University Press)

Lewes, George Henry (1850). *The Biographical History of Philosophy from its Origin in Greece down to the Present Day* (NY: D. Appleton)

Lewes, George Henry (1853). *Comte's Philosophy of the Sciences: Being an Exposition of the Principles of the Cours de Philosophie Positive of Auguste Comte* (London: H.G. Bohn)

Lewes, George Henry (1874-75). *Problems of Life and Mind*, first series, 2 vols. (London: Truebner and Co.)

Lewes, George Henry (1877). *Problems of Life and Mind*, second series, or *The Physical Basis of Mind*, (London: Truebner and Co.)

Lewes, George Henry (1879). *Problems of Life and Mind*, third series, 2 vols. (London: Truebner and Co.)

Lewontin, Richard C., Steven Rose, and Leon Kamin (1984). *Not in our Genes: Biology, Ideology, and Human Nature* (NY: Pantheon Books)

Limoges, Camille (1970). *La Sélection Naturelle; études sur la première constitution d'un concept, 1837-59* (Paris: PUF)

Lloyd, Alfred H. (1927). "Also the Emergence of Matter", *J. Phil.* 24: 309-332

Lloyd Morgan, Conwy (1882). *Water and Its Teachings, in Chemistry, Physics, and Physiography: A Suggestive Handbook* (London: Edward Stanford)

Lloyd Morgan, Conwy (1884a). *Facts Around Us: Simple Readings in Inorganic Science; with Experiments* (London: Edward Stanford)

Lloyd Morgan, Conwy (1884b). "Instinct", *Nature*, 29: 370-374

Lloyd Morgan, Conwy (1885). *Springs of Conduct: An Essay in Evolution* (London: Kegan Paul)

Lloyd Morgan, Conwy (1887). *Animal Biology: An Elementary Textbook* (London: Rivingtons, 4th edition revised, 1911)

Lloyd Morgan, Conwy (1888). "Elimination and Selection", *Bristol Naturalists' Soc. Proc.* 5: 1-13

Lloyd Morgan, Conwy (1890). *Animal Life and Intelligence* (London: Arnold)

Lloyd Morgan, Conwy (1891). *Animal Sketches* (London: Arnold)

Lloyd Morgan, Conwy (1892a). "Mental Evolution: An Old Speculation in a New Light", *Monist* 2: 161-177

Lloyd Morgan, Conwy (1892b). "The Law of Psychogenesis", *Mind*, NS 1: 72-93

Lloyd Morgan, Conwy (1893). "The Doctrine of Auta", *Monist*, 3: 161-175

Lloyd Morgan, Conwy (1894a). *An Introduction to Comparative Psychology* (London: The Walter Scott Pub. Co. Ltd.), 2nd edition revised, 1903

Lloyd Morgan, Conwy (1894b). *Psychology for Teachers* (London: Arnold), 2nd edition rewritten, 1906

Lloyd Morgan, Conwy (1894c). "Instinct and Intelligence in Chicks and Ducklings", *Natural Science* 4: 204-213

Lloyd Morgan, Conwy (1894d). "Three Aspects of Monism", *Monist* 4: 321-332

Lloyd Morgan, Conwy (1894e). "Weismann on Heredity and Progress", *Monist*, 4: 17-23

Lloyd Morgan, Conwy (1895). "A Piece of Patchwork", *Monist*, 5: 354-362

Lloyd Morgan, Conwy (1896a). "Animal Automatism and Consciousness", *Monist* 7: 1-18 and 8: 148-149

Lloyd Morgan, Conwy (1896b). *Habit and Instinct* (London: Arnold)

Lloyd Morgan, Conwy (1896c). "Naturalism", *Monist* 6: 76-90

Lloyd Morgan, Conwy (1897). "The Realities of Experience", *Monist* 8: 1-18

Lloyd Morgan, Conwy (1898a). "Causation: Physical and Metaphysical", *Monist* 8: 230-249

Lloyd Morgan, Conwy (1898b). "The Philosophy of Evolution", *Monist* 8: 481-501

Lloyd Morgan, Conwy (1898c). "Instinct and Intelligence in Animals", *Royal Institution of Great Britain* reprint

Lloyd Morgan, Conwy (1898d). "Animal Intelligence", *Natural Science* 13: 265-272)

Lloyd Morgan, Conwy (1899a). "Vitalism", *Monist* 9: 177-196

Lloyd Morgan, Conwy (1899b). "Biology and Metaphysics", *Monist* 9: 538-562

Lloyd Morgan, Conwy (1900a). *Animal Behavior* (London: Arnold)

Lloyd Morgan, Conwy (1900b). "Psychology and the Ego", *Monist* 10: 62-84

Lloyd Morgan, Conwy (1900c). "The Conditions of Human Progress", *Monist* 10: 422-440

Lloyd Morgan, Conwy (1905). *The Interpretation of Nature*. The Lowell Lectures for 1905 (Bristol: Arrowsmith).

Lloyd Morgan, Conwy (1907). "Organic Selection", *Proc. Cotteswold Club* 16: 65-69

Lloyd Morgan, Conwy (1909a). "Mental Factors in Evolution", in Seward, A. C. (ed.): *Darwin and Modern Science*, pp. 424-445

Lloyd Morgan, Conwy (1909b). "The Natural History of Experience", *Br. J. of Psych.* 3: 1-20

Lloyd Morgan, Conwy (1910). "Instinct and Intelligence", *Br. J. Psych.* 3: 219-229

Lloyd Morgan, Conwy (1912a). *Instinct and Experience* (London: Methuen)

Lloyd Morgan, Conwy (1912b). "Creative Evolution", notes dated Feb. 24, 1912. *Lloyd Morgan Archives* DM612, 4 pp.

Lloyd Morgan, Conwy (1912c). "Some Aspects of M. Bergson's Philosophy", notes dated 15 March 1912. *Lloyd Morgan Archives* DM612, 5pp.

Lloyd Morgan, Conwy (1912d). "Scientific Thought; being Notes for a Course of Lectures to Senior Science Students", *Lloyd Morgan Archives* 128/553, 219 pp.

Lloyd Morgan, Conwy (1913a). "M. Bergson's Outline of Instinct", notes for a talk to the Cardiff Phil. Soc, April 28, 1913 *Lloyd Morgan Archives* DM612, 4 pp.

Lloyd Morgan, Conwy (1913b). *Spencer's Philosophy of Evolution*. The Spencer Lecture for 1913 (Oxford: Clarendon Press)

Lloyd Morgan, Conwy (1914). "Notes on Berkeley's Doctrine of Esse", *Proc. Aristot. Soc.* 15: 100-139

Lloyd Morgan, Conwy (1915). "Mind and Body in their Relation to Each Other and to External Things", *Scientia* 18: 244- 256

Lloyd Morgan, Conwy (1916). "Fact and Truth", *Proc. Aristot. Soc.* 17: 195-215

Lloyd Morgan, Conwy (1917). "Enjoyment and Awareness", *Mind* 101: 1-11

Lloyd Morgan, Conwy (1919). *Eugenics and Environment* (London: John Bale)

Lloyd Morgan, Conwy (1921a). "Consciousness and the Unconscious" [Presidential Address to Section J (Psychology) of the British Association for the Advancement of Science] in *Report of the Eighty Ninth Meeting, BAAS* at Edinburgh, 1921 (London: John Murray, 1922)

Lloyd Morgan, Conwy (1921b). "Instinctive Behaviour and Enjoyment", *B. J. Psych*, 12: 1-30

Lloyd Morgan, Conwy (1921c). "Huxley Lecture", notes dated Nov. 25, 1921, in *Lloyd Morgan Archives* DM612,14 pp.

Lloyd Morgan, Conwy (1922). "Foreword" to Bianchi, Leonardo (1922). *The Mechanism of the Brain and the Function of the Frontal Lobes,* tr. James H. Macdonald (Edinburgh: E. and S. Livingston) pp. 11-15

Lloyd Morgan, Conwy (1923). *Emergent Evolution.* The Gifford Lectures for 1922 (London: Williams and Norgate)

Lloyd Morgan, Conwy (1924a). "A Philosophy of Evolution", in Muirhead, J. H. (ed.): *Contemporary British Philosophy: Personal Statements* ,First Series (London: George Allen and Unwin), pp. 272-306

Lloyd Morgan, Conwy (1924b). "Relation between Physical and Psychical Nexus", in *Proc. Aristot. Soc*, supp. vol. 4: 162-169

Lloyd Morgan, Conwy (1925a). "Biology", in James Jeans, Harold Jeffreys et al: *Evolution in the Light of Modern Knowledge: A Collective Work* (London: Blackie and Son)

Lloyd Morgan, Conwy (1925b). "Naturalisme et Vie", *Scientia* 38: 7-18

Lloyd Morgan, Conwy (1926a). "Influence and Reference: A Biological Approach to Philosophical Problems", *Monist* 36: 535-560

Lloyd Morgan, Conwy (1926b). *Life, Mind and Spirit.* The Gifford Lectures for 1923 (London: Williams and Norgate)

Lloyd Morgan, Conwy (1926c). "Objects Under Reference" (Presidential Address to the Aristotelian Society), in *Proc. Aristot. Soc.* 27: 1-20

Lloyd Morgan, Conwy (1927a). "A Concept of the Organism, Emergent and Resultant", *Proc. Aristot. Soc*, 27: 141- 176

Lloyd Morgan, Conwy (1927b). "Review of Holism and Evolution", *J. Phil. Studies* 2: 93-97

Lloyd Morgan, Conwy (1928). "Mind in Evolution", in Mason, Francis (ed.): *Creation by Evolution: A Consensus* (NY: The Macmillan Co.), pp. 340 - 354

Lloyd Morgan, Conwy (1929a). "Emergence" in *Encyclopaedia Britannica*, 14th edition, v.8, pp. 390-391

Lloyd Morgan, Conwy (1929b). "Evolution and Mind", in *Encyclopaedia Britannica*, 14th edition, v.8, pp. 929- 931

Lloyd Morgan, Conwy (1929c). *Mind at the Crossways* (London: Williams and Norgate)

Lloyd Morgan, Conwy (1929d). "The Case for Emergent Evolution", *J. Phil. Studies* 4: 23-38

Lloyd Morgan, Conwy (1929e). "The Case for Emergent Evolution", letter to the editor, *J. Phil. Stud.* 4: 431- 432

Lloyd Morgan, Conwy (1930a). *The Animal Mind* (London: Arnold)

Lloyd Morgan, Conwy (1930b). "The Bifurcation of Nature", *Monist* 40: 161-181

Lloyd Morgan, Conwy (1931). "A Psychological Approach to Space-Time", *Mind* 40: 409-423

Lloyd Morgan, Conwy (1932a). "Autobiography", in Murchison, Karl (ed.): *A History of Psychology in Autobiography* (Worcester, Mass.: Clark University Press), pp. 237-264

Lloyd Morgan, Conwy (1932b). "Evolution and Creative Purpose", manuscript in *Lloyd Morgan Archives*, typescript, 437 pp.

Lloyd Morgan, Conwy (1933). *The Emergence of Novelty* (New York: Henry Holt and Co.)

Lloyd Morgan, Conwy (1934a). "Psychology and Beyond" in *Science for a New World: The Scientific Outlook on World Problems Explained by Leading Exponents of Modern Scientific Thought*, planned and arranged by the late J. Arthur Thomson, ed. J. G. Crowther.

Lloyd Morgan, Conwy (1934b). "The Ascent of Mind", in *The Great Design*, ed. Frances Baker Mason (NY: The Macmillan Co.), pp. 115-132

Lovejoy, Arthur O. (1926). "The Meanings of 'Emergence' and Its Modes", in Brightman, Edgar S. (ed.) *Proceedings of the Sixth International Congress of Philosophy* (NY: Longmans Green and Co.), pp. 20-33

Lovejoy, Arthur O. (1936). *The Great Chain of Being: The Study in the History of an Idea* (Cambridge, Mass.: Harvard Univ. Press)

Lowry, Ann. (1974): "A Note on Emergence", *Mind*, 83: 276-77

Lyell, Charles (1830-33). *Principles of Geology, or, the Modern Changes of the Earth and Its Inhabitants, Considered as Illustrative of Geology* (first edition 1830-33 in 4 volumes London: John Murray, 12th edition, 1875)

Lyell, Charles (1861). *The Geological Evidences of the Antiquity of Man, with Remarks on Theories of the Origin of Species by Variation* (Philadelphia: George W. Childs)

Lyell, Charles, (1873) *The Geological Evidences of the Antiquity of Man, with an Outline of Glacial and Post-Tertiary Geology and Remarks on the Origin of*

Species, with Special Reference to Man's First Appearance on the Earth (London: J. Murray)

Lyell, Charles (1970). *Sir Charles Lyell's Scientific Journals on the Species Questions,* edited by Leonard G. Wilson (New Haven: Yale University Press).

Mackay, D. M. (1978). "Selves and Brains", *Neurosci.* 3: 599-606

Mackenzie, W. Leslie (1926). "The Notion of Emergence", *Aristot. Soc. Supp.* Vol. 6: 56-68 [Symposium on Emergence]

Malisoff, William Marias (1939). "Emergence Without Mystery", *Phil. Sc.* 6: 17-24

Malthus, Thomas Robert (1803) *An Essay on the Principle of Population; or, A View of its Past and Present Effects on Human Happiness; With an Inquiry into our Prospects Respecting the Future Removal or Mitigation of the Evils which it Occasions*, new edition (London: T. Bensley)

Manier, E. (1978). *The Young Darwin and His Cultural Circle* (Dordrecht: D. Reidel Pub. Co.)

Margulis, Lynn (1970). *Origin of Eukaryotic Cells: Evidence and Research Implications for a Theory of the Origin and Evolution of Microbial, Plant, and Animal Cells on the Precambrian Earth* (New Haven, CT: Yale Univ. Press)

Marvin, Walter (1912). *A First Book in Metaphysics* (NY: The Macmillan Company)

Mayr, Ernst (1960). "The Emergence of Evolutionary Novelties", in *Evolution After Darwin*, ed. Sol Tax, vol. 3 (Chicago: U. of Chicago Press)

Mayr, Ernst (1961). "Cause and Effect in Biology", in Mayr, Ernst (1976). *Evolution and the Diversity of Life, Selected Essays* (Cambridge, Mass: Harvard, The Belknap Press)

Mayr, Ernst (1976). *Evolution and the Diversity of Life, Selected Essays* (Cambridge, Mass.: Harvard, The Belknap Press)

Mayr, Ernst (1982). *The Growth of Biological Thought: Diversity, Evolution and Inheritance* (Cambridge, Mass.: Harvard, The Belknap Press)

Mayr, Ernst (1987). "The Ontological Status of Species: Scientific Progress and Philosophical Terminology", *Bio. and Phil.* 2: 145-166

Mayr, Ernst, (1988). "Is Biology an Autonomous Science", in Mayr, Ernst (1988), *Towards a New Philosophy of Biology* (Cambridge, Mass: The Belknap Press of Harvard Univ. Press)

McDougall, William (1901). "On the Seat of the Psycho-Physical Processes", *Brain*, 24: 577-630

McDougall, William (1905). *Physiological Psychology* (London: J. Dent and Son)

McDougall, William (1913). *Body and Mind: A History and a Defense of Animism* (NY: Macmillan and Co.)

McDougall, William (1925). "Mental Evolution", in *Evolution in the Light of Modern Knowledge*, ed. James Jeans et al. (London: Blackie and Son)

McDougall, William (1929). *Modern Materialism and Emergent Evolution* (London: Methuen and Co.)

McDougall, William (1930). "Autobiography", in *A History of Psychology in Autobiography*, vol. 1, ed. Carl Murchison (Worcester, Mass: Clark University Press)

McDougall, William (1938). *Riddle of Life: A Survey of Theories* (London: Methuen and Co.)

Medawar, Peter (1969). *Induction and Intuition in Scientific Thought* (Oxford: Oxford Univ. Press)

Medawar, Peter (1974). "A Geometric Model of Reduction and Emergence", in Ayala, Francisco and Th. Dobzhansky (1974). *Studies in the Philosophy of Biology: Reductionism and Related Problems* (Berkeley: Univ. of California Press)

Mead, George Herbert (1932). *The Philosophy of the Present*, ed. A. E. Murphy (Chicago: Univ. of Chicago Press)

Mead, George Herbert (1934). *Mind, Self and Society from the Standpoint of a Social Behaviorist*, ed. C. W. Morris (Chicago: Univ. of Chicago Press)

Mead, George Herbert (1938). *The Philosophy of the Act*, ed. A. E. Murphy (Chicago: Univ. of Chicago Press)

Meehl, P. E. and Sellars, Wilfrid (1956). "The Concept of Emergence", in *Minnesota Studies in the Philosophy of Science*, ed. Herbert Feigl and Michael Scriven, vol. 1 (Minneapolis: U. of Minnesota Press)

Metz, Rudolf (1938). *A Hundred Years of British Philosophy*, trans. by J. W. Harvey, T. E. Jessop, and Henry Sturt, edited by J. H. Muirhead (London, G. Allen & Unwin Ltd.)

Mill, J. S. (1872). *A System of Logic Ratiocinative and Inductive, Being a Connected View of the Principles of Evidence and the Methods of Scientific Investigation*, 4th ed. (London: John W. Parker and Son)

Mivart, St. George (1871). *On the Genesis of Species* (NY: D. Appleton and Co.)

Mivart, St. George (1889). *On the Origin of Human Reason, being an Examination of Recent Hypotheses Concerning it* (London: Kegan, Paul, Trench and Co.)

Montagu, Ashley (1952). *Darwin, Competition and Cooperation* (New York, H. Schuman)

Montague, William Pepperell (1929). "A Materialistic Theory of Emergent Evolution", in Felix Adler, E. A. Ames et al. (eds.) *Essays in Honour of John Dewey* (NY: Henry Holt and Co.), pp. 257-273

Montague, William Pepperell (1931). "Confessions of an Animistic Materialist", in Adams, George P. and Montague, William Pepperell (eds.), *Contemporary American Philosophy: Personal Statements*, vol. II (London: George Allen and Unwin Ltd.), pp. 134-159

Morgan, Thomas Hunt (1932). *The Scientific Basis of Evolution* (New York, W. W. Norton & Company)

Morris, C. R. (1926). "The Notion of Emergence", *Aristot. Soc. Supp.* Vol. 6: 49-55 [Symposium on Emergence]

Nabours, Robert K. (1930). "A Third Alternative: Emergent Evolution", *Sc. Monthly* 31: 453-456

Nagel, Ernest (1961). *The Structure of Science Problems in the Logic of Scientific Explanation* (NY: Harcourt, Brace & World)

Needham, Joseph (1928). "Organicism in Biology", *J. Phil.* 3: 29-40

Needham, Joseph (1936). *Order and Life* (New Haven: Yale University Press)

Needham, Joseph (1937). *Integrative Levels: A Reevaluation of the Idea of Progress.* The Herbert Spencer Lectures. (Oxford: Clarendon Press)

Needham, Joseph (1942). *Time, The Refreshing River* (NY: Macmillan and Co.)

Needham, Joseph (1956). *Science and Civilization in China*, vol. 2: *History of Scientific Thought* (Cambridge: University Press)

Nernst, Walther (1895). *Theoretical Chemistry from the Standpoint of Avogadro's Rule and Thermodynamics*, revised in accordance with the 6th German ed. by H.T. Tizard (London: Macmillan, 1911)

Neurath, Otto, Rudolf Carnap and Charles Morris, eds. (1970). *Foundations of the Unity of Science: Toward an International Encyclopaedia of Unified Science*, 2 vols. (Chicago, Univ. of Chicago Press)

Newton, Isaac (1729). *Mathematical Principles of Natural Philosophy, and his System of the World*, tr. Andrew Motte, rev. Florian Cajori, 2 vols. (Berkeley: Univ. of California Press, 1962)

Novikoff, Alex. B. (1945). "The Concept of Integrative Levels and Biology", *Science* 101: 209-215

Oparin, A. I. (1938). *The Origin of Life* (New York, The Macmillan Company)

Osborn, Henry Fairfield (1926). *Evolution and Religion in Education: Polemics of the Fundamentalist Controversy of 1922 to 1926* (NY: Charles Scribner's Sons)

Ostwald, Wilhelm (1890). *Outlines of General Chemistry*, tr. by James Walker (London: Macmillan and Co.)

Paley, William (1807). *Natural Theology; or, Evidences of the Existence and Attributes of the Deity. Collected from the Appearances of Nature*, 11th ed. (London: R. Faulder)

Pap, Arthur. (1952). "The Concept of Absolute Emergence", *Br. J. Phil. Sc.* 2: 302-311

Parsons, J. H. (1936). "Conwy Lloyd Morgan: 1852-1936", *Obit. Notices of Fellows of the Royal Soc*, 2: 24-27

Patrick, G. T. W. (1922)." The Emergent Theory of Mind", *J. Phil.* 19: 701-708

Patrick, G. T. W. (1926). "Review of Lloyd Morgan's Life, Mind and Spirit", *J. Phil.* 23: 554-557

Pepper, Stephen C. (1926). "Emergence", *J. Phil.* 23: 241-245

Polanyi, Michael. (1958) *Personal Knowledge: Towards a Post-Critical Philosophy* (N.Y: Harper and Row, 1964)

Popper, Karl (1957) *The Poverty of Historicism* (Boston, Beacon Press)

Popper, Karl (1963). *Conjectures and Refutations: The Growth of Scientific Knowledge* (NY: Harper Torchbooks)

Popper, Karl (1972). *Objective Knowledge: An Evolutionary Approach* (Oxford: Oxford Univ. Press)

Popper, Karl (1974). "Scientific Reduction and the Essential Incompleteness of All Science", in Ayala, Francisco and Th. Dobzhansky (1974). *Studies in the Philosophy of Biology: Reductionism and Related Problems* (Berkeley: Univ. of California Press), also in Popper, Karl (1982). *The Open Universe: An Argument for Indeterminsm*, ed. W. W. Bartley, III (Totowa, NJ: Rowman and Littlfield), pp. 131-162

Popper, Karl (1978). "Natural Selection and the Emergence of Mind", *Dialectica* 32: 339-355

Popper, Karl and John Eccles (1977). *The Self and Its Brain* (Cambridge: Cambridge Univ. Press)

Powell, Baden (1855). *Essays on the Spirit of the Inductive Philosophy, the Unity of Worlds, and the Philosophy of Creation* (London: Longman, Brown, Green and Longmans)

Priestley, Joseph (1776). *Disquisitions Relating to Matter and Spirit* (New York: Arno Press, 1975)

Prigogine, Ilya and Isabelle Stengers (1984). *Order out of Chaos: Man's New Dialogue with Nature* (Boulder, Colorado: New Science Library)

Ralston, Harold Jameson. *Emergent Evolution and Purpose: The Concept of Purpose in the Philosophy of C. Lloyd Morgan* (Iowa: State University of Iowa)

Raven, Charles E. (1962) *Teilhard de Chardin: Scientist and Seer* (NY: Harper & Row)

Ray, John (1691). *The Wisdom of God as Manifested in the Works of the Creation*, reprinted (NY: Garland Publishing, 1978)

Redfield, R. (1942), ed. *Levels of Integration in Biological and Social Systems*, in *Biol. Sympos. 8* (Lancaster: J. Cattell Press)

Reiser, Oliver L. (1924a). "Creative Monism", *Monist* 34: 395-425

Reiser, Oliver L. (1924b). "Life as a Form of Chemical Behavior", *Monist* 34: 150-160

Reiser, Oliver L. (1925). "An Electromagnetic Theory of Matter, Life, and Mind', *Monist* 35: 605-632

Reiser, Oliver L. (1926). "Probability, Natural Law, and Emergence", *J. Phil.* 23: 421-435

Reiser, Oliver L. (1930). "Mathematics and Emergent Evolution", *Monist* 40: 509-525

Reiser, Oliver L. (1935). *Philosophy and the Concepts of Modern Science* (NY: The Macmillan Co.)

Rhodes, Frank H. T. (1987). "Darwinian Gradualism and Its Limits: The Development of Darwin's Views on the Rate and Pattern of Evolutionary Change", *J. Hist. Bio.* 20: 139-157

Richards, Robert J. (1987). *Darwin and the Emergence of Evolutionary Theories of Mind and Behavior* (Chicago: Univ. of Chicago Press)

Rignano, Eugenio. "Entre le Vitalism et le Mécanisme s'impose une solution intermédiaire, *Scientia* 38: 253-272, 335-362

Ritter, W. E. and J. Bailey (1928). "The Organismal Conception", *Univ. of California Publications in Zoology* 31: 307-355

Romanes, George John (1883a). *Animal Intelligence* (NY: D. Appleton and Co.)

Romanes, George John (1883b). *Mental Evolution in Animals* (London: Kegan, Paul and Trench)

Romanes, George John (1888). *Mental Evolution in Man* (London: Kegan Paul and Trench)

Romanes, George John (1892-97). *Darwin and After Darwin: An Exposition of the Darwinian Theory and Discussion of Post-Darwinian Questions*, ed. by C. Lloyd Morgan 3 vols. (London: Longmans, Green and Co.)
v.1. *The Darwinian Theory*
v.2. *Post-Darwinian Questions: Heredity and Utility*
v.3. *Post-Darwinian Questions: Isolation and Physiological Selection*

Romanes, George John (1895). *Mind and Motion and Monism*, ed. C. Lloyd Morgan (NY: Longmans, Green and Co.)

Ruse, Michael (1970). "The Revolution in Biology", *Theoria* 10: 13-22

Ruse, Michael (1979). *The Darwinian Revolution: Science Red in Tooth and Claw* (Chicago: Univ. of Chicago Press)

Russel, E. S (1926). "The Notion of Emergence", *Aristot. Soc. Supp.* Vol. 6: 39-48 [Symposium on Emergence]

Russell, Bertrand (1914). *Our Knowledge of the External World* (NY: W. W. Norton and Co, 1929)

Russell, Bertrand (1917). *Mysticism and Logic* (London: Longmans, Green and Co.)

Russell, Bertrand (1927a). *Philosophy* (NY: W. W. Norton and Company)

Russell, Bertrand (1927b). *The Analysis of Matter* (London: Allen and Unwin)

Russell, Bertrand (1931). *The Scientific Outlook* (NY: W. W. Norton and Co, 1962)

Salk, Jonas (1983). *Anatomy of Reality: Merging of Intuition and Reason* (NY: Columbia Univ. Press)

Salthe, Stanley (1985). *Evolving Hierarchical Systems: Their Structure and Representation* (NY: Columbia University Press)

Schievella, Pat S. (1967). *The Philosophy of Conwy Lloyd Morgan* (Unpublished doctoral dissertation: Columbia University)

Schneirla, T. C. (1949). "A Biological Survey of Integrative Levels", in Sellars, Roy Wood, V. J. McGill and Marvin Farber (eds.), *Philosophy for the Future: The Quest for Modern Materialism* (NY: The Macmillan Company)

Schneirla, T. C. (1951). "The 'Levels' Concept in the Study of Social Organization in Animals", in *Selected Writings of T. C. Schneirla* (1972), ed. Lester R. Aronson, Ethel Tobach, Jay S. Rosenblatt and Daniel S. Lehrman (San Francisco: W. H. Freeman and Company)

Schilpp, Paul Arthur (1963) *The Philosophy of Rudolf Carnap*, The Library of Living Philosophers, v.11 (La Salle, Ill., Open Court)

Schwartz, Joel S. (1984). "Darwin, Wallace and the Descent of Man", *J. Hist. Bio.* 17: 271-289

Sellars, Roy Wood (1922). *Evolutionary Naturalism* (Chicago: Open Court Co.)

Sellars, Roy Wood (1926). *The Principles and Problems of Philosophy* (NY: Macmillan Co.)

Sellars, Roy Wood (1927). "Why Naturalism and Not Materialism?", *Phil. Rev.* 36: 216-225

Sellars, Roy Wood (1933). "L'Hypothèse de l'émergence", *Rev. Meta. et Morale* 40: 309-324

Sellars, Roy Wood (1944a). "Is Naturalism Enough?", *J. Phil.* 41: 533-544

Sellars, Roy Wood (1944b)."Can a Reformed Materialism Do Justice to Values?", *Ethics* 55: 28-45

Sellars, Roy Wood (1944c). "Reformed Materialism and Intrinsic Duration", *Phil. Rev.* 53: 359-382

Sellars, Roy Wood (1949), ed., with V. J. McGill and Marvin Farber. *Philosophy for the Future: The Quest for Modern Materialism* (NY: The Macmillan Company)

Sellars, Roy Wood (1951). "Professor Goudge's Queries with Respect to Materialism", *Phil. Rev.* 60: 243-248

Sellars, Roy Wood (1959). "Levels of Causality: The Emergence of Guidance and Reason in Nature", *Phil. Phenom. Res.* 20: 1-17

Sellars, Roy Wood (1960). "Panpsychism or Evolutionary Materialism", *Phil. Sc.* 27: 329-350

Sellars, Roy Wood (1970). *The Principles, Perspectives and Problems of Philosophy* (NY: Pageant Press Int. Corp.), 2nd edition of *Principles and Problems of Philosophy* (1926)

Sellars, Wilfrid and Meehl, P. E. (1956). "The Concept of Emergence", in Feigl, Herbert and Scriven, Michael (eds.): *Minnesota Studies in the Philosophy of Science*, vol. 1: *The Foundations of Science and the Concepts of Psychology and Psychoanalysis* (Minneapolis: University of Minnesota Press), pp. 239-252

Simpson, George Gaylord (1944). *Tempo and Mode in Evolution* (NY: Columbia Univ. Press)

Simpson, George Gaylord, ed. (1982). *The Book of Darwin* (NY: Washington Square Press)

Smart, J. J. C. (1981). "Physicalism and Emergence", *Neurosci.* 6: 109-113

Smuts, Jan (1926). *Holism and Evolution* (NY: Macmillan and Co.)

Spaulding, Edward Gleason (1912). "A Defense of Analysis", in *The New Realism: Cooperative Studies in Philosophy* (1912) by Edwin B. Holt, Walter T.Marvin, William Pepperell Montague, Ralph Barton Perry, Walter B. Pitkin, and Edward Gleason Spaulding (New York, Macmillan)

Spaulding, Edward Gleason (1918). *The New Rationalism: The Development of a Constructive Realism Upon the Basis of Modern Logic and Science Through the Criticism of Opposed Philosophical Systems* (NY: Henry Holt and Co.)

Spencer, Herbert (1852). "The Development Hypothesis", in *Essays: Scientific, Political, and Speculative*, vol. 1 (NY: D. Appleton and Company, 1910), pp. 1-7

Spencer, Herbert (1857). "Progress: Its Law and Cause", in *Essays: Scientific, Political, and Speculative*, vol. 1 (NY: D. Appleton and Company, 1910), pp. 8-62

Spencer, Herbert (1879). *The Data of Ethics* (NY: P. F. Collier and Son, 1902)

Spencer, Herbert (1885). *The Principles of Sociology*, vol. 1 (NY: D. Appleton and Co., 1910)

Spencer, Herbert (1886). "The Factors of Organic Evolution", in *Essays: Scientific, Political, and Speculative*, vol. 1 (NY: D. Appleton and Company, 1910), pp. 389-466

Spencer, Herbert (1893-1902). *A System of Philosophy* , also issued as *Synthetic Philosophy* (NY: D. Appleton and Co, 1910)
v.1. *First Principles of a New System of Philosophy*
v.2-3. *Principles of Biology*
v.4-5. *Principles of Psychology*
v.6-8. *Principles of Sociology*
v.9-10. *Principles of Ethics*

Sperry, Roger W. (1952). "Neurology and the Mind-Brain Problem", *Am. Sci.* 291-312

Sperry, Roger W. (1964). "Problems Outstanding in the Evolution of Brain Function", James Arthur Lecture on the Evolution of the Human Brain, (NY: The American Museum of Natural History)

Sperry, Roger W. (1969). "A Modified Concept of Consciousness", *Psych. Rev.* 76: 532-536

Sperry, Roger W. (1970). "An Objective Approach to Subjective Experience: Further Explanation of an Hypothesis", *Psych. Rev.* 77: 584-590

Sperry, Roger W. (1980). "Mind-Brain Interaction: Mentalism, Yes; Dualism, No", *Neurosci.* 5: 195-206

Sperry, Roger W. (1981). "Changing Priorities", *Ann. Rev. Neurosci.* 4: 1-15

Srinivasiengau, K. R. (1934). "Emergent Evolution: An Indian View", *Phil. Rev.* 43: 598-606

Stace, W. T. (1939). "Novelty, Indeterminism, and Emergence", *Phil. Rev.* 48: 296-310

Sumner, Francis B. (1929). "Is Evolution a Continuous or Discontinuous Process?", *Sc. Monthly* 29: 72-78

Teilhard de Chardin, Pierre. (1959) *The Phenomenon of Man*, with an intro. by Julian Huxley, tr. Bernard Wall (NY: Harper)

Thomson, J. Arthur (1920). *The System of Animate Nature*, 2 vols. The Gifford Lectures for 1915 and 1916. (London: Holt)

Thomson, J. Arthur (1925). *Concerning Evolution.* The Terry Lectures (Cambridge: Harvard University Press)

Tully, R. E. (1981). "Emergence Revisited", in Sumner, L. W. (ed): *Pragmatism and Purpose: Essays Presented to Thomas A. Goudge* (Toronto: University of Toronto Press), pp. 261-277

Wallace, Alfred Russel (1870). *Contributions to the Theory of Natural Selection: A Series of Essays* (London: Macmillan and Co.)

Wallace, Alfred Russel (1874). "A Defence of Modern Spiritualism", *Fortnightly Review* 21: 630-657, 785-807

Wallace, Alfred Russel (1889). *Darwinism, An Exposition of the Theory of Natural Selection,m with some of its Implications* (London: Macmillan and Co.)

Wallace, Alfred Russel (1905). *My life: A Record of Events and Opinions* (New York, Dodd, Mead & Company)

Wallace, Alfred Russel (1911). *The World of Life, A Manifestation of Creative Power, Directed Mind and Ultimate Purpose* (London: Chapman and Hall)

Weismann, August (1889). *Essays upon heredity and kindred biological problems*, Ed. by Edward B. Poulton, Selmar Schonland and Arthur E.Shipley (Oxford: Clarendon press)

Weismann, August (1893). *The Germ-Plasm: a Theory of Heredity*, tr. by W. Newton Parker and Harriet Ronnfeldt (New York: Scribner's)

Weiss, Paul (1969). "The Living System: Determinism Stratified", in Koestler, Arthur and J. R. Smythies, eds., *Beyond Reductionism: New Perspectives in the Life Sciences* (NY: Macmillan)

Wells, F. L. (1928). "Psychogenic Factors in Emergentism and Allied Views", *J. Phil.* 25: 71-75

Wheeler, W. M. (1926). "Emergent Evolution of the Social", in Brightman, Edgar S. (ed.) *Proceedings of the Sixth International Congress of Philosophy* (NY: Longmans, Green and Co.), pp. 33-45, also reprinted in *Science*, Nov. 5, 1926, LXIV: 433-440

Wheeler, W. M. (1928). *Emergent Evolution and the Development of Societies* (NY: W. W. Norton and Co.)

Whitehead, Alfred North (1925) *Science and the Modern World*, Lowell lectures for 1925 (NY: The Macmillan Company)

Whitehead, Alfred North (1926). *Religion in the Making*, Lowell lectures for 1926 (NY: The Macmillan Company)

Whitehead, A. N. (1929). *Process and Reality*, Gifford Lectures for 1927-28 (Cambridge: University Press)

Whitehead, A. N. (1933). *Adventures of Ideas* (Cambridge: University Press)

Wilson, Donna (1969). "Forms of Hierarchy: A Selected Bibliography". *Gen. Syst.* 14: 3-15

Wilson, E. O. (1971). *The Insect Societies* (Cambridge, Mass: Harvard Univ. Press)

Wilson, E. O. (1975). *Sociobiology: The New Synthesis* (Cambridge, Mass: Harvard Univ. Press)

Wood-Jones, F. (1942). *Design and Purpose* (London: Macmillan)

Woodger, J. H. (1929). *Biological Principles: A Critical Study* (London: Routledge and Kegan Paul)

Woodger, J. H. (1930-31). "The 'Concept of Organicism' and the Relation between Embryology and Genetics", *Quart. Rev. Bio.* 5: 1-21, 438-463, 6: 178-207

Woodger, J. H. (1937). *The Axiomatic Method in Biology* (Cambridge: University Press)

Woodger, J. H. (1952). *Biology and Language* (Cambridge: Univ. Press)

Yolton, John W. (1983). *Thinking Matter: Materialism in Eighteenth-Century Britain* (Minneapolis: University of Minnesota Press)

INDEX OF NAMES

INDEX OF SUBJECTS

Episteme

A SERIES IN THE FOUNDATIONAL,
METHODOLOGICAL, PHILOSOPHICAL, PSYCHOLOGICAL, SOCIOLOGICAL, AND
POLITICAL ASPECTS OF THE SCIENCES, PURE AND APPLIED

---

1. W.E. Hartnett (ed.): *Foundations of Coding Theory*. 1974 ISBN 90-277-0536-4
2. J.M. Dunn and G. Epstein (eds.): *Modern Uses of Multiple-Valued Logic*. 1977 ISBN 90-277-0747-2
3. W.E. Hartnett (ed.): *Systems: Approaches, Theories, Applications*. 1977 ISBN 90-277-0822-3
4. W. Krajewski: *Correspondence Principle and Growth of Science*. 1977 ISBN 90-277-0770-7
5. J.L. Lopes and M. Paty (eds.): *Quantum Mechanics, a Half Century Later*. 1977 ISBN 90-277-0784-7
6. H. Margenau: *Physics and Philosophy*. Selected Essays. 1978 ISBN 90-277-0901-7
7. R. Torretti: *Philosophy of Geometry from Riemann to Poincaré*. 1978 ISBN Hb 90-277-0920-3 / Pb 90-277-1837-7
8. M. Ruse: *Sociobiology: Sense or Nonsense?* 1979; 2nd ed. 1985 ISBN Hb 90-277-1797-4 / Pb 90-277-1798-2
9. M. Bunge: *Scientific Materialism*. 1981 ISBN 90-277-1304-9
10. S. Restivo: *The Social Relations of Physics, Mysticism, and Mathematics*. Studies in Social Structure, Interests, and Ideas. 1983 ISBN Hb 90-277-1536-X / Pb (1985) 90-277-2084-3
11. J. Agassi: *Technology*. Philosophical and Social Aspects. 1985 ISBN Hb 90-277-2044-4 / Pb 90-277-2045-2
12. R. Tuomela: *Science, Action, and Reality*. 1985 ISBN 90-277-2098-3
13. N. Rescher: *Forbidden Knowledge* and Other Essays on the Philosophy of Cognition. 1987 ISBN 90-277-2410-5
14. N.J. Moutafakis: *The Logics of Preference*. A Study of Prohairetic Logics in Twentieth Century Philosophy. 1987 ISBN 90-277-2591-8
15. N. Laor and J. Agassi: *Diagnosis: Philosophical and Medical Perspectives*. 1990 ISBN 0-7923-0845-X
16. F.P. Ramsey: *On Truth*. Original Manuscript Materials (1927–1929) from the Ramsey Collection at the University of Pittsburgh, edited by N. Rescher and U. Majer. 1991 ISBN 0-7923-0857-3
17. H.A. Shenkin: *Medical Ethics*. Evolution, Rights and the Physician. 1991 ISBN 0-7923-1031-4
18. E. Agazzi (ed.): *The Problem of Reductionism in Science*. 1991 ISBN 0-7923-1406-9
19. D. Blitz: *Emergent Evolution*. Qualitative Novelty and the Levels of Reality. 1992 ISBN 0-7923-1658-4

KLUWER ACADEMIC PUBLISHERS – DORDRECHT / BOSTON / LONDON